The Cambridge Manuals of Science and
Literature

LINKS WITH THE PAST IN
THE PLANT WORLD

Sequoia magnifica Knowlton, in the Yellowstone National Park.
(From a photograph kindly supplied by Prof. Knowlton.) See p. 104.

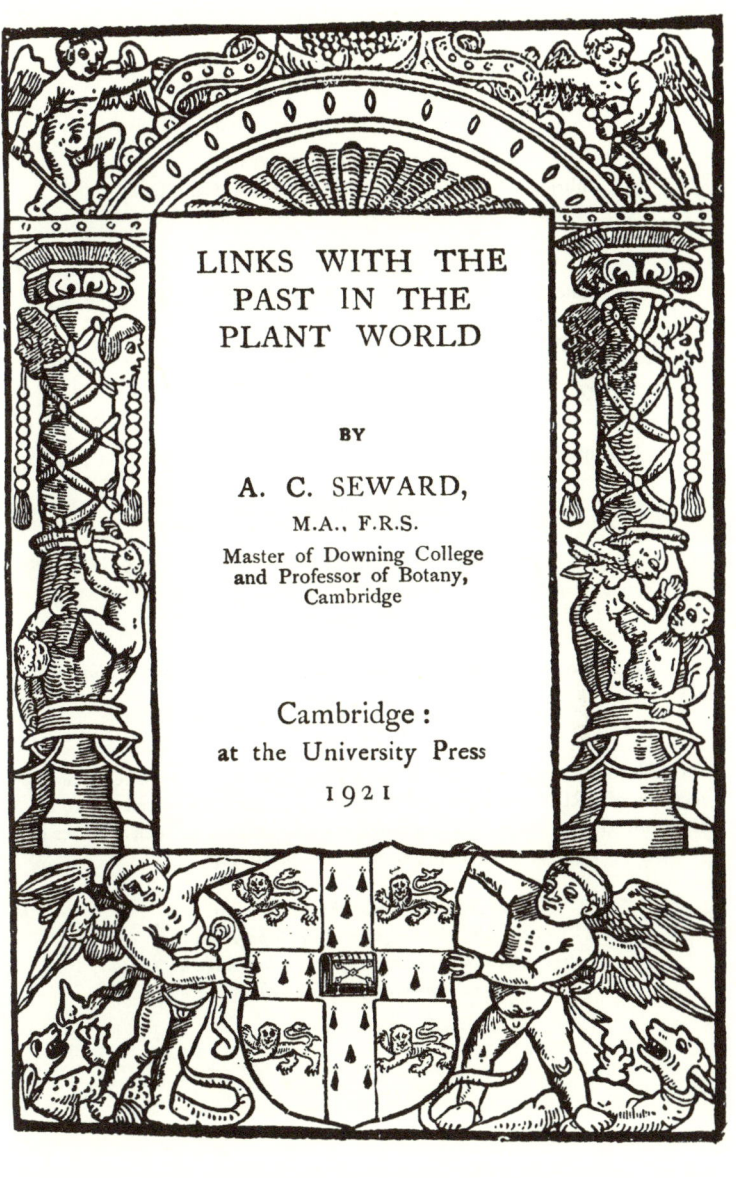

LINKS WITH THE PAST IN THE PLANT WORLD

BY

A. C. SEWARD,

M.A., F.R.S.

Master of Downing College
and Professor of Botany,
Cambridge

Cambridge :
at the University Press

1921

CAMBRIDGE UNIVERSITY PRESS
Cambridge, New York, Melbourne, Madrid, Cape Town,
Singapore, São Paulo, Delhi, Tokyo, Mexico City

Cambridge University Press
The Edinburgh Building, Cambridge CB2 8RU, UK

Published in the United States of America by
Cambridge University Press, New York

www.cambridge.org
Information on this title: www.cambridge.org/9781107401594

First published 1911
Reprinted 1921
First paperback edition 2011

A catalogue record for this publication is available from the British Library

ISBN 978-1-107-40159-4 Paperback

*With the exception of the coat of arms
at the foot, the design on the title page is a
reproduction of one used by the earliest known
Cambridge printer, John Siberch, 1521*

PREFACE

MY object in writing this book is primarily to call attention to some of the many questions which are raised by an enquiry into the relative antiquity of existing plants, and to illustrate the nature of the evidence afforded by the records of the rocks. One may agree with the dictum, 'There is but one art—to omit,' but to practise this art is often a difficult task. While fully conscious of the incompleteness of the treatment of the subjects dealt with in these pages, and of defects in the method of presentation, I hope that I may succeed in attracting some of my readers who are already interested in living plants to the study of plants of former ages.

I am greatly indebted to my colleague Dr C. E. Moss for reading the proofs and for many valuable suggestions. I wish to thank Mr and Mrs Clement Reid, Prof. MacDougal of the Arizona Desert Laboratory, Prof. Campbell of Stanford University, Prof. F. H. Knowlton of the United States National Museum, Washington, Mr A. G. Tansley, Prof. Yapp, and Mr W. R. Welch for photographs which they have allowed me to reproduce. As on many previous occasions, I am indebted to my wife for contributing drawings.

A. C. SEWARD.

Botany School, Cambridge.
July 1911.

The numbers in brackets interspersed in the text refer to the Bibliography at the end of the volume.

CONTENTS

CHAPTER I

INTRODUCTORY : THE LONGEVITY OF TREES, ETC.

'Believe me who have tried. Thou wilt find something more in woods than in books. Trees and rocks will teach what thou canst not hear from a master.' ST BERNARD.

The recent publication in the daily press of instances of human longevity under the heading 'Links with the Past' prompted a comparison between the length of time represented by the duration of a tree and the lifetime of a human being. The comparison of single lives suggested the further step of contrasting the antiquity of the oldest family-histories with the remoteness of the period to which it is possible to trace the ancestry of existing members of the plant kingdom.

My primary object in these pages is not to deal with familiar cases of longevity in trees, but to consider in the first place some of the problems connected with the origin of the present British flora, and then to describe a few examples of different types of plants whose ancestors flourished during periods of the earth's history long ages before the advent of the human race.

In dealing with plants of former ages we are confronted with the difficulty of forming an adequate conception of the length of time embraced by geological periods in comparison with the duration of the historic era. Some of the 'Selections from the Greek Papyri' recently edited by Dr Milligan (Cambridge 1910) refer to common-place events in terms familiar to us in modern letters : we forget the interval of 2000 years which has elapsed since they were written. Similarly, the close agreement between existing plants and species which lived in remote epochs speaks of continuity through the ages, and bridges across an extent of time too great to be expressed by ordinary standards of measurement. Terms of years when extended beyond the limits to which our minds are accustomed cease to have any definite meaning. While there is a certain academic interest in discussions as to the age of the earth as expressed in years, we are utterly unable to realise the significance of the chronology employed. After speaking of the futility of attempting to introduce chronological precision into periods so recent as those which come into the purview of archaeologists, Mr Rice Holmes suggests a method better adapted to our powers. He says—'Ascend

the hill on which stands Dover Castle, and gaze
upon Cape Grisnez, let the waters beneath you
disappear ; across the chalk that once spanned the
channel like a bridge men walked from the white
cliff that marks the horizon to where you stand.
No arithmetical chronology can spur the imagination
to flights like these(1).' On the other hand, the
use in some country districts in Britain of spindles
almost identical with instruments used in spinning
by the ancient Egyptians, and similar survivals
described by the author of a book entitled *The
Past in the Present*(2), bring within the range
of our vision an early phase of the historic era.
The rude implements still fashioned by the flint-
knappers of Brandon in Suffolk connect the present
with the Palaeolithic age. Measured from the
standpoint of historic reckoning, survivals from
prehistoric days appeal to us as persistent types
which have remained unchanged in a constantly
changing world.

In one of his essays Weismann quotes an old
German saying with regard to comparative longevity,
which asserts that 'a wren lives three years, a dog
three times as long as a wren' and so on in a regu-
larly ascending series ; the life of a deer is estimated
at three times that of a crow and an oak three times
that of a deer, which means that, computed on this
basis, an oak lives nearly 20,000 years(3) ! This

fanciful illustration of the relative longevity of an
oak is the expression of a truth, namely the superiority
of trees over animals in regard to the duration of
life.   As a seventeenth-century translator of Pliny's
Natural History writes, 'In old times trees were the
very temples of the gods : and according to that
antient manner, the plaine and simple peasants of
the country, savouring still of antiquity, do at this
day consecrate to one God or other, the goodliest
and fairest trees that they can meet withal.'   Oaks
growing in Pliny's day in the Hercynian forest are
said to have been there 'ever since the creation of
the world(4).'   Sir Joseph Hooker, in an account
of some Palestine oaks, gives a drawing of a famous
tree at Mamre, known as Abraham's Oak, which is
supposed to mark the spot where the Patriarch
pitched his tent (5).   Examples such as these,
though of no scientific value, serve to illustrate the
well-founded belief in the extraordinary longevity of
trees.   In the absence of evidence to the contrary, it
would be rash to deny the possibilty that William
the Conqueror's Oak in Windsor Forest, described
by Loudon in his *Arboretum Britannicum* and
mentioned by later writers, may be a survival from
the reign of the king whose name it bears.   Although
it is seldom possible to state with confidence the
exact age of old oaks and yews famed for length of
days, there can be no doubt as to the enormous

antiquity of many of our trees whose years are
'sacred with many a mystery.' The section of a
trunk of one of the mammoth trees of California
(*Sequoia gigantea*) exhibited in the Natural History
department of the British Museum, shows on its
polished surface 1335 concentric rings denoting
successive increments of wood produced by the
activity of a cylinder of cells situated between the
hard woody tissue and the bark. It is generally
assumed that each year a tree produces a single
ring, though, as is well known, an estimate of age
calculated on this assumption cannot be regarded as
more than an approximation to the truth. If this
giant tree, which was felled in 1890, was then 1335
years old, it had already reached an age of over two
centuries when Charlemagne was crowned Emperor
at Rome. The concentric rings on a tree trunk owe
their existence to certain structural differences between the wood formed in the spring and in the late
summer. In Sequoia, as in other members of the
great class of cone-bearing trees, the wood is composed of comparatively narrow elements which serve
to carry water from the roots to the branches and
leaves. As spring succeeds winter the inactivity
of the plant-machine is followed by a period of
energetic life ; opening buds and elongating shoots
create a demand for a plentiful supply of ascending
sap, and in reponse to this the tree produces a fresh

cylinder of wood composed of relatively wide conducting tubes.   After the first rush of life is succeeded
by a phase of more uniform and gentler activity, the
demand for water becomes less exacting and the
wood which is formed during the rest of the growing
season consists of narrower water-pipes.   A period of
rest ensues, until in the following spring new layers
of larger tubes are laid down in juxtaposition to the
narrower elements of the latest phase of the preceding
summer.   This alternation of larger and smaller
tubes produces the appearance of concentric rings on
a cross-section of a tree.   It is not the pause in the
active life of the plant which is responsible for the
effect of rings, but the fact that the wood produced
immediately before and immediately after the pause
is not structurally identical.   In trees grown under
the more uniform conditions of certain tropical
regions, the annual rings are either feebly developed or
absent ; for example, in some Indian oaks the wood
shows no concentric rings of growth.

Stated in general terms, rings of growth in the
wood of a tree are the expression of a power possessed
by the plant of regulating the structure of its component elements in response to the varying nature of
the external stimuli.   In certain circumstances, for
example after the destruction of the young buds by
caterpillars, the tree makes a special effort to repair
the loss by producing a new set of shoots.   This may

be recorded by the occurrence of two concentric rings
in one season. An extreme instance of departure
from the normal has recently been described(6) in
which a tree of *Theobroma cacao* (the cocoa tree),
planted in Ceylon in the summer of 1893 and felled
in January 1901, after a life of just over 7 years, was
found to have 22 rings in its stem. In this case the
tree shed its leaves three times a year, and each break
in the uniformity of its vital activities was registered
by the apposition of what under ordinary conditions
are spoken of as spring and late-summer wood.
At Aden trees stated by natives to be very old
showed only five or six rings of wood, a fact connected
with the almost complete lack of rain and with the
uniform conditions of existence.

The degree of accuracy to be allowed to estimates
of age founded on the number of 'annual' rings is,
however, of secondary importance in comparison
with the enormously greater hold on life possessed
by trees as contrasted with the higher animals. Early
in the nineteenth century the swiss botanist A. P.
de Candolle expressed the opinion that trees do not
die from senile decay, but only as the result of injury
or disease. Trees are constructed on a plan funda-
mentally different from that underlying the structure
of the highly complex human organism, and are thus
endowed with a sort of potential immortality. Between a coral-reef and a tree there are many essential

differences, but a rough analogy may be recognised. It has been suggested that some of the large corals in the Red Sea which are still tenanted by living polyps may have been growing in the days of the Pharaohs. The polyp represents the growing portion of a lifeless mass of coral rock which is constantly extended by the activity of the organism at the summit of each branch. A tree, unlike the higher animals, does not reach a stage at which the whole of its substance attains a condition of permanence and fixity. It consists of a complex branching-system in which each shoot increases in length by virtue of the youthful vigour of its apex : to a large extent the tree as a whole consists of lifeless material incapable of further growth, as is the case of the older portions of a coral-reef ; but the regular increase in girth of the trunk and its branches demonstrates that this comparison is only partially true, and that the power of growth in a tree is not confined to the extremities of the youngest shoots. The tip of every twig is composed of minute cells endowed with a potentiality of development like that which characterises the embryonic tissues of a seedling just emerged from the seed. In the course of its growth, each branch, by means of its living and dividing cells, contributes to the several parts of the complex mechanism of the tree. While the greater number of cells acquire a permanent form and lose the

power of further development, there remains a cylinder of cells endowed with perpetual youth. This
cylinder of living cells, known as the cambium, extends
between the wood and bark from one end of the tree
to the other ; by its periodic activity it adds new layers
of tissue each year and thus, by increasing the amount
of conducting tubes for the transport of water and
for the distribution of elaborated food, it enables the
tree to respond to the increasing demands which
are the necessary accompaniment of increasing size.
It has already been pointed out that in the spring
when the sap flows most vigorously the cambial
cylinder produces larger tubes and afterwards, when
the tree settles down to its normal life, these are
succeeded by narrower and stronger tubes. These
later formed elements serve also an important me-
chanical purpose ; by the strength of their walls they
increase the supporting power of the tree and enable
it to sustain the added burden of the annual increase
in the weight and extent of its spreading branches.

It is the persistence of permanently juvenile tissue
in certain regions of a tree, together with the remarkable power of repairing injuries and shedding
effete parts, that constitute some of the most striking
contrasts between the higher animals and plants.
The embryo oak in the earlier stages of development
consists entirely of actively growing cells ; by degrees
differentiation of the embryonic tissues results in the

localisation of regions of cell-production at the tips
of the elongating stem and root.   These apical groups
of cells are, in fact, portions of the embryonic organism
which persist so long as the plant lives.   This continuity between the growing tip of an old oak stem
and the cells of the undifferentiated embyro affords
one of the most remarkable examples in nature of
a link between the past and the present.

If we pass beyond the stretch of time represented
by the life of a single tree, a few successive generations suffice to carry our retrospect back to the days
when forests of oaks, birches, and other trees im-
peded the progress of the Roman invaders, and, a
stage farther back, to the age of Neolithic man whose
remains are occasionally found in our heaths and
moors and in the submerged forests round our coast.
The blocks of oak and beech, some of which are as
sound as when first felled, recently discovered below
the foundations of parts of Winchester Cathedral constructed at the end of the twelfth or in the opening
years of the thirteenth century, are relics of Norman
forests.   In the course of some excavations at Brigg in
Lincolnshire in 1886 a dug-out boat was found nearly
50 feet long and from 4 to 5 feet in breadth.   The stem
of the oak from which the canoe had been fashioned
shows no sign of branching for a length of over 40
feet, a fact which points to the growth of the tree
in a forest where the race for light induced the

development of clean columnar stems. The Brigg 'dug-out,' now in the Hull Museum, was discovered in an old alluvial valley of the Ancholme river, formerly connected with the Humber, and it may be that it was used by Neolithic man as a ferry for river-service[7].

From the period claimed by archaeologists we pass by gradual stages into the domain of the geologist. As Huxley wrote, 'when even the dim light of Archaeology fades, there yet remains Palae-ontology, which...has brought to daylight once more the exuvia of ancient populations, whose world was not our world, who have been buried in river beds immemorially dry, or carried by the rush of waters into caves, inaccessible to inundation since the dawn of traditions[8].' The length of time represented by a succession of long-lived individuals of the same species becomes enormously extended when we pass to the history of families, and disinter from the sediments of other ages the remains of extinct types. As we descend the geological series familiar types gradually disappear, and through a succession of changing floras we penetrate to the fragmentary records contained in the older rocks until the absence of documents sets a limit to our quest.

The Scots pine shares with the oak, the beech, the aspen, the yew, and several other trees the right to be included in the native flora of Britain. In the peat-beds of Scotland even up to 3000 feet above sea-level the stumps of pines occur in abundance, and in many places recent researches have revealed the occurrence of successive forests of pines, oaks, and spruces separated from one another by the

(A)

Fig. 1. *Pinus sylvestris* Linn. in the Black Wood of Rannoch. (Photograph by Mr A. G. Tansley.)

accumulations of swampy vegetation[9]. The spruce fir has long ceased to be a member of the British flora, but in a few localities in the Scottish Highlands patches of primeval pine forests remain. The

accompanying photograph (Fig. 1), taken by my friend
Mr A. G. Tansley, in the Black Wood of Rannoch in
north-west Perthshire, shows a few trees of *Pinus
sylvestris* growing in their native soil : the form of
the older tree (*A*) suggests comparison with that of
a well-grown beech such as we are familiar with
in English plantations.   This spreading dome-shaped
habit seems to be a peculiarity of the Highland tree,
and is one of the characters which have led some
botanists to regard it as a variety (*Pinus sylvestris*
var. *scotica*) of the ordinary Scots pine.   Though it
is doubtful if any relics of primeval pine woods are
left in England, abundant evidence of the former
existence of the Scots pine is afforded by the submerged forests exposed at low-tide on many parts of
the English and Welsh coasts and at the base of
some of the English peat moors.   During the con-
struction of the Barry docks on the north coast of
the Bristol Channel a few years ago, the exposed
sections of peat and forest beds were investigated by
Dr Strahan and by Mr Clement Reid.   There is
evidence of a subsidence of the land to an extent of
55 feet since the formation of the lower peat-beds
containing oak, hazel, willow, and other trees.   The
pine, unknown in Wales during the historic period,
was recognized in the Barry cutting.   The occurrence
of a polished flint implement assigns a date to the
uppermost portion of this old land-surface[10].

It is impossible within the limits of a small volume to discuss in detail the evidence furnished by the records of the rocks as to the relative antiquity of the different constituents of the present vegetation of Britain. In later chapters a few selected plants are described which are pre-eminently ancient types. Before passing to the consideration of the data on which the geological history of plants is based, brief reference may be made to one of the most interesting and difficult problems of botanical research, namely the history of the British flora.

CHAPTER II

THE GEOGRAPHICAL DISTRIBUTION OF PLANTS

'No speculation is idle or fruitless that is not opposed to truth
or to probability, and which, whilst it co-ordinates a body of well
established facts, does so without violence to nature, and with a due
regard to the possible results of future discoveries.'

SIR JOSEPH HOOKER.

In the vegetation of the British Isles the leading
rôle is played by that large group to which the term
Flowering Plants is frequently applied. This group,
including the two sub-divisions Dicotyledons and
Monocotyledons, is known by the name Angiosperms,
a designation denoting the important fact that the
seeds are developed in an ovary or protective seed-
case (ἀγγεῖον, a vessel or box). The fact that these
highly elaborated products of development made
their appearance, so far as we know, at a comparatively late stage in the history of the plant-world,
attests their efficiency as a class and demonstrates
the rapidity with which they have overspread the
surface of the earth as successful competitors in

the struggle for existence. As Darwin wrote in a letter to Sir Joseph Hooker in 1881, 'Nothing is more extraordinary in the history of the vegetable kingdom, as it seems to me, than the *apparently* very sudden or abrupt development of the higher plants(11).' In another letter (1879) to the same friend we read, 'The rapid development as far as we can judge of all the higher plants within recent geological times is an abominable mystery(12).' Making allowance for the probability, or indeed certainty, that the imperfection of the geological record tends to exaggerate the apparent suddenness of the appearance of this vigorous class, and allowing for the fact that our knowledge of the records of the rocks in which the highest plants first occur is very incomplete, we cannot escape from the conclusion that this recently evolved group spread with amazing rapidity. Various reasons may be suggested in explanation of the dominant position which the Angiosperms hold in the floras of the world. As an instance of their rapid increase during the Cretaceous epoch[1], the period which has furnished the earliest satisfactory records of Flowering Plants, the following statement by an American writer may be quoted :—'The rapidity with which it [*i.e.* the group of Flowering Plants] advanced, conquering or

[1] For the position of the Cretaceous and other systems in the geological series, see the table on page 42.

supplanting all rivals, may be better understood when we remember that it forms 85% of the flora of the Dakota group'; that is a series of sedimentary rocks in Dakota referred by geologists to the middle of the Cretaceous period[13]. In the Wealden rocks of England, which are rich in the remains of Lower Cretaceous plants, no undoubted Flowering Plant has so far been found.

The more efficient protection of the ovules, the germs which, after fertilisation, become the seeds, the extraordinary variety in the floral mechanisms for assisting cross-pollination, the arrangements for nursing the embryo, and the structural features of the wood in relation both to rapid transport of water and to the storage of food, are factors which have probably contributed to the success of the Angiosperms. The degree of weight to assign to each contributing cause cannot as yet be satisfactorily determined, but the general question raised by the recent origin of these latest products of evolution offers a promising field for work. While admitting our inability at present to do more than suggest possibilities, we may encourage research by speculation.

The members of the Vegetable Kingdom placed next to the Flowering Plants are the Gymnosperms or naked-seeded (γυμνός, naked) plants, including (i) the Conifers, *e.g.* pines, firs, larches, the yew, etc.,

(ii) a small group of plants known as the Cycads,
whose existing members, now almost confined to
a few tropical regions, are the descendants of a
vigorous race represented by many species in the
floras of the Mesozoic epoch. A third sub-division
of the Gymnosperms, the Ginkgoales, is represented
by a single survivor, which is described in a later
chapter as one of the most remarkable links with the
past in the plant kingdom.

The Gymnosperms are geologically very much
older than the Angiosperms. Members of this class
played a prominent part in the vegetation of the
Coal age and it is certain that they existed in the
still older Devonian period. The only other group
to which reference is made in later chapters is that
of the Ferns, one of the sub-divisions of a large class
known as the Vascular Cryptogams or Pteridophyta.
These plants, like the Gymnosperms, are represented
in the oldest floras of which recognisable remains
have been preserved. The main groups of the
vegetable kingdom, founded on existing plants, are
distinguished by well-defined differences ; they are
comparable with separate twigs of a tree springing
from larger branches and these again uniting below
in a common trunk. The vegetation of to-day represents only the terminal portions of the upper
branches. As we descend the geological series,
records of extinct types are found which enable

us either to trace the separate branches to a common origin or to recognise a convergence towards a common stock. Were a botanist to find himself in a forest of the Coal age he would experience great difficulty in assigning some of the plants to their systematic position: characters now regarded as distinguishing features of distinct groups would be met with in combination in a single individual. It is by the discovery of such generalised types, which serve as finger-posts pointing the way to lines of evolution, that the student of pre-existing plants has been able to throw light on the relative antiquity of existing forms, and to trace towards a common ancestry plants which now show but little indication of consanguinity.

Confining our attention to the dominant group of plants in the British flora, namely the Flowering Plants, we may profitably consider the question, though we cannot satisfactorily answer it,—which members of this group are entitled to be regarded as the most ancient inhabitants? The past history of our native plants, and their geographical range, not only in the British Isles but on the Continent of Europe, are subjects well worthy of the attention of field-botanists whose interests are apt to be confined within too narrow bounds. There are numerous problems relating to the composition of the present vegetation of Britain which might be discussed in reference to the relative antiquity of plants; but in a single chapter it is impossible to do more than call attention to certain considerations which are frequently overlooked by students of British species.

It is customary to speak of the British flora as consisting for the most part of species introduced into this country by natural means, while some plants owe their introduction to human agency or are 'escapes' from cultivation. It is by no means an easy task in some instances to decide whether a species is native or introduced, but in some cases, a few of which are mentioned, there is no doubt as to the alien nature of the plants. The term 'native' needs a word of explanation. It is not intended to convey the idea that a plant so designated came into existence on British soil and spread thence to other regions; but by native species we mean such as have reached this country by migration from other lands, or it may be in some instances have actually originated in this part of Europe. One of the best known aliens in Britain is the American water-weed, *Elodea canadensis* (or *Anacharis alsinastrum*), which was discovered about sixty years ago in a canal near Market Harborough in Leicestershire (14). In all probability this North American species was introduced into England with timber. Once established, it spread through the waterways with alarming rapidity and became a serious pest. Elodea affords an admirable instance of the serious interference with the balance of Nature by the introduction of a new competitor into an environment conducive to vigorous development. Another foreign water-plant, *Naias graminea*, for the importation of which Egyptian cotton may be responsible, has been recorded from the Reddish canal near Manchester (15). This African and Asiatic species occurs in Europe only as a colonist; it is said to have been introduced into Italy with East Indian rice. A more recent case of alien immigration due to unintentional human agency is that of *Potamogeton pennsylvanicus*, a pond-weed of Canada, the United States, Jamaica, and elsewhere. Specimens of this species were first noticed in 1907 in a canal near Halifax close to the effluent from a cotton mill. Since its discovery the plant has slightly extended its range. It is suggested by Mr Bennett, who first identified the alien, that its fruits were brought to this country in goods from the United States (16).

Of the introduction of these and other foreign plants we have satisfactory records; but there are many others which may owe their presence to man's agency, though we have no information as to their arrival.

It has long been recognised that several members of the British flora are related to Scandinavian species. The Scandinavian flora, as Sir Joseph Hooker says in his well-known paper on the ' Outlines of the Distribution of Arctic plants,' not only girdles the globe in the Arctic Circle, and dominates over all others in the North Temperate Zone of the Old World, but intrudes conspicuously into every other temperate flora, whether in the northern or southern hemisphere, or on the Alps of tropical countries'(17). The view generally held is that during the Glacial period this Arctic flora was driven South, and aided by land-bridges, which were afterwards submerged, many of the northern migrants found a more congenial home in Britain. It is however by no means improbable that this conclusion may have to be considerably modified. Mr and Mrs Reid, as the result of their careful analysis of the Pre-Glacial Flora of Britain, express the opinion that 'the pre-glacial plants suggest climatic conditions almost identical with those now existing, though slightly warmer' (27, 2). It is noteworthy that the list of plants given in their paper does not include any typical Arctic species. The occurrence on the mountains of Scotland and elsewhere of such plants as *Silene acaulis, Dryas octopetala, Saxifraga oppositifolia* and other Saxifrages *Rubus chamaemorus* (the Cloudberry), and the dwarf Birch illustrate the Arctic-Alpine element in our flora.

The opinion is held by many Swiss botanists that

their Alpine species have in large measure been derived from non-glaciated parts of the Pyrenees, that is from a region which was presumably able to retain its flora at a time when more northern lands were exposed to extreme Arctic conditions. My friend Dr Moss believes that some of the so-called Scandinavian plants came to Britain from Central Europe after the retreat of the ice ; if this view is correct it means that some at least of our Arctic-Alpine plants reached these islands by a southern rather than by a northern route.

Interesting examples of far-travelled northern plants recently described by Professor Engler of Berlin afford additional illustrations of the general principles enunciated many years ago by Sir Joseph Hooker. A species of flowering Rush, *Luzula spicata* var. *simensis*, occurs at an altitude of 3600 metres in Abyssinia and on Kilimanjaro. *Luzula spicata* is found in the whole of the Arctic and Subarctic belt in Scotland, Auvergne, the Jura mountains, and elsewhere. The species probably began its career in the northern hemisphere where it grew abundantly on the higher ground in the Arctic Circle : it eventually travelled along the North American Andes and appeared in Mexico under a guise sufficiently distinct to warrant the use of another name, *Luzula racemosa*. In an eastern direction it reached the Himalayas and is

represented in Abyssinia by a closely allied form.
From Abyssinia to Kilimanjaro *Luzula spicata* 'had
to travel a long distance ; but it is not impossible
that it either still exists or has existed previously
on a few of the high mountains between Abyssinia
and Kenia, from which, having advanced to the
Kilimanjaro, it again produced new forms....At any
rate, it is impossible to do without distribution of
seeds of alpine plants by air-currents or by birds
from one mountain to the other in explaining the
history of distribution'[18].

The majority of British plants are identical with
species in Central and Northern Europe : of these,
some are among the most widely spread English
species, *e.g.* the Daisy and Primrose, while others,
such as the Oxlip (*Primula elatior*), are confined to
the Eastern counties, and others, such as the Cheddar
Pink (*Dianthus caesius*), are restricted to Western
counties.

Before considering a small section of the British
flora which is the most interesting from the point of
view of origin, a short digression may be allowed in
order to call attention to the importance of a branch
of science which Darwin spoke of as 'that grand
subject, that almost keystone of the laws of creation,
geographical distribution,' and in 1847 referred to
as 'that noble subject of which we as yet but dimly
see the full bearing.' It was largely as the results

of his study of distribution in the Galapagos Islands that Darwin determined to 'collect blindly every sort of fact which bears any way on what are species.' The acceptance of the view 'that each species was first produced within a single region'(19), raises the subject of geographical distribution to a far higher plane than it occupied in pre-Darwinian days. Although most people are familiar with some of the commoner means by which plants are able to colonise new ground through the adaptation of their fruits and seeds to various methods of transport, the conception of a plant as a stationary organism tends to prevent due allowance being made for the comparative facility with which, in the course of successive generations, a species is able to wander from place to place. The individual animal is endowed with powers of locomotion enabling it to seek new feeding grounds and to avoid enemies; but with the exception of some of the simplest forms a plant cannot move—'le matin la laisse où la trouve le soir.'

The rate of travel may or may not be rapid, but in a comparatively short time, if the conditions are favourable, a tree may spread over a wide area. Mr Ridley, Director of the Botanic Gardens, Singapore, writes as follows in reference to the rate of travel of one of the common Malayan trees (*Shorea leprosula*), which bears winged fruits particularly well adapted to wind-transport : ' If we assume that a tree flowers and fruits at 30 years of age and the fruits are disseminated to a distance of 100 yards, that the furthest fruits always germinate and so continue in one direction, it will be seen that under such most favourable circumstances the species can only spread 300 yards in 100 years, and would take 58,666 years to migrate 100 miles '[20].

There is, however, one type of distribution—what is called discontinuous distribution—to which special attention should be directed on account of its intimate association with questions relating to the past history of living organisms. Many examples might be quoted from both the animal and plant kingdoms in support of the view that discontinuous distribution is a criterion of antiquity. When identical or very nearly identical plants occur in regions separated from one another by areas in which the particular species is unknown, the inference is either that the surviving individuals are remnants of a large number formerly distributed over a wider continuous area, or that in the course of evolution similar conditions in widely separated areas led to the production of identical types. The former view is much the more probable : it is consistent with the conclusions arrived at on other grounds as to the connexion between discontinuous distribution and ancient lineage. The explanations of the widespread occurrence among

different races of similar objects or legends afford an
analogous case. As Dr Andrew Lang points out in
Custom and Myth, it is held by some students that
the use of the bull roarer—to cite a specific instance
—by different peoples denotes descent from a common
stock, though he considers the more probable explanation to be that similar minds, working with
simple means towards similar ends, might evolve the
bull roarer and its mystic uses anywhere.

The Cedars of Lebanon afford an interesting example of discontinuous distribution. They illustrate
how a species, which may be assumed to have
originated in one region, in the course of its
wanderings may undergo slight changes until, at
a later stage when the plants have disappeared from
parts of the once-continuous area, the remaining
outlying groups of individuals are spoken of under
different specific names. The cedars of Lebanon,
known as *Cedrus libani*, occur as isolated groups
on the Lebanon hills as outliers of the larger forests
of the Taunus 250 miles distant. The African cedar,
Cedrus atlantica, is separated from the Lebanon
cedar by a distance of 1400 miles. Approximately
the same distance divides the Lebanon cedar from
the deodar, *Cedrus deodara*, which extends from
Afghanistan along the Himalayas almost to the
confines of Nepal. Sir Joseph Hooker regards the
three cedars as varieties of one species which once

formed a continuous forest: he attributes the present discontinuous distribution, in part at least, to the effects of a warmer succeeding a colder climate. The less favourable conditions drove the vegetation of the lowlands to seek more congenial habitats at higher altitudes. In this connexion it is interesting to find that in Algeria the cedar is confined to the higher ground where the snow lies long in the spring(21).

The Tulip Tree of North America and Central China affords one of many examples of existing flowering plants which illustrate the close connexion between present distribution and past history. The genus Liriodendron, often cultivated in the south of England, is now represented by two species, the best known of which—the Tulip Tree, *Liriodendron tulipifera*—extends from Vermont to Florida and westwards to Lake Michigan and Arkansas. The leaves bear a superficial resemblance to those of the Sycamore, but are as a rule easily distinguished by the truncated form of the apex; the specific name was suggested by the tulip-like form of the flowers. Fossil leaves of Liriodendron are not uncommon in the Cretaceous rocks of Disco Island in latitude 70° N., where they occur with other flowering plants which bear striking testimony to the mildness of the Cretaceous climate in high northern latitudes. One of the associated flowering plants is a species of

Artocarpus, described by Dr Nathorst as *Artocarpus Dicksoni* which bears a close resemblance to *A. incisa* the bread-fruit tree of the southern tropics of the Old World. Without attempting to deal fully with the past history of Liriodendron, it may be confidently stated that the records of the rocks are consistent with the idea of antiquity suggested by the present distribution of the two surviving species.

Islands such as Great Britain and Ireland, situated a short distance from a continent, contain many plants which are widely spread in different parts of the world, together with a very small number peculiar to the British Isles though closely allied to species on the neighbouring continent or to plants farther afield. The occasional recognition of species previously believed to be confined to Britain tends to reduce the short list of our endemic types.

An enquiry into the origin of an island flora involves a consideration of the data in regard to changes in level and relative distribution of land and water in the course of geological evolution. It is generally agreed that at no distant date, in a geological sense, Great Britain and Ireland were united to the continent. There is, however, another fact to reckon with, namely the prevalence of Arctic conditions in northern Europe when a thick sheet of ice spread over the greater part of the British Isles. There can be no doubt that the severity of the climate

during the Glacial period was such as to destroy
a large proportion of the vegetation. The question
is, were all the flowering plants destroyed or were
some of the hardier species able to survive, either on
the higher peaks which kept their heads above the
level of the ice or on the southern fringe of England
beyond the ice-covered region? It is impossible to
give a definite answer : the probability is that nearly
all the pre-Glacial species were destroyed, but it is
not impossible that some Alpine-Arctic plants escaped
extinction, while others retreated to more southern
and less Arctic areas by means of a land-connexion
with France or crossed the intervening sea by ocean-
currents, by animal agency, or by wind.

Although we possess but imperfect information as
to the extent and duration of land-bridges between
Britain and the continent, there are no special
difficulties in the way of accounting for the presence
of Scandinavian, Germanic, and other elements in the
British flora. There are, however, other and more
difficult problems to consider in reference to a small
group of flowering plants which are met with in the
west and south of Ireland, also, to a less extent, in
Cornwall and in a few other localities in the south-
west of England. In Connemara in the west of Ireland,
where hard frosts are unknown and winter snows are
rare, there are three kinds of Heath, St Dabeoc's Heath
(*Daboecia polifolia*), the Mediterranean Heath (*Erica*

mediterranea) and *Erica Mackaii* which are not
found elsewhere in the British Isles or in the whole
of northern Europe, but reappear in the Pyrenees.
The London Pride (*Saxifraga umbrosa*), another
Pyrenean plant, grows on the south and west coast
of Ireland from Waterford to Donegal. *Arbutus
Unedo*, the Strawberry tree, which flourishes in the
Killarney district of County Kerry and occurs in
neighbouring localities, has a wide distribution in
the Mediterranean region.   Devonshire and Cornwall
possess two other Heaths, *Erica ciliaris*, which extends into Dorsetshire and occurs in north Brittany,
and *Erica vagans*, both Pyrenean species, while a
Mediterranean plant, *Gladiolus illyricus*, grows in the
New Forest.

In 1846 Edward Forbes dealt with the problems
presented by the distribution of British plants in an
essay which has exercised a far-reaching influence.
When Forbes published his work, comparatively little
was known as to the possibilities of transport of seeds
and fruits across barriers of water[22].   His conviction
that the known means of dispersal were insufficient
to account for the presence of Mediterranean or
Lusitanian plants in Ireland led him to turn to
geology for a solution of the problem.   He was thus
led to put forward the view that in the course of
the Tertiary period when, as we know from palaeontological evidence, the climate of north and west

Europe was much warmer than it is now, and long before the beginning of the climatic changes which culminated in the Glacial period, there was a land-connexion between the west of Ireland and the south-west of the continent. Mr Praeger, whose work on the Irish flora is well known to systematic botanists, agrees with the conclusions of Forbes, and sees in the Portuguese and Mediterranean plants 'relics of a vegetation which once spread along a bygone European coast-line which stretched unbroken from Ireland to Spain'[23]. If this explanation is correct it entitles Arbutus, St Dabeoc's heath and other members of this southern group to be regarded as a very old section of our flora. There is, however, another side to the question : granting that a certain number of Irish plants were able to withstand the rigours of an Ice age, it is hardly likely that the strawberry tree and other southern types, which it is admitted flourish in the south-west of Ireland because of the mildness of the climate, were of the number of those which endured an extreme Arctic phase. Moreover, if these Mediterranean species are survivals from the Tertiary period, if they have been isolated since pre-Glacial days as an outlier of a southern flora, we might fairly expect that during the long interval between their arrival and the present day new forms would have been produced closely related to, though not identical with, the

parent types.   This, however, has not been proved to
be the case.   Darwin in speaking of Forbes' Essay in
a letter to de Candolle in 1863 says that he differs
from most of his contemporaries 'in thinking that
the vast continental extensions of Forbes, Heer, and
others are not only advanced without sufficient evidence, but are opposed to much weighty evidence'(12).
The alternative view is to regard Arbutus and its
compatriots as post-Glacial arrivals and not as survivals from a widely spread Tertiary flora.

A recently published account of the New Flora
of the volcanic island of Krakatau furnishes an
instructive and remarkable demonstration of the
facility with which a completely sterilised island,
separated by several miles of ocean from neighbouring lands, may be restocked with vegetation(24).
In 1883 the island of Krakatau, then densely covered
with a luxuriant tropical vegetation, was partially
destroyed by a series of exceptionally violent volcanic
explosions.   After this catastrophe only a third of
the island was left : the surface was deeply covered
by pumice and volcanic ash and no vestige of life
remained.   In 1906 a party of botanists who spent
a few hours on Krakatau collected 137 species of
plants : the vegetation was in places so dense that
it was with the greatest difficulty they penetrated
beyond the coastal belt, and some of the trees had
reached a height of 50 feet.   The seeds and fruits

of this new flora have been carried by ocean-currents, by wind, and by the agency of birds from other islands in the Malay Archipelago. The nearest islands, except the small island of Sebesi, about 12 miles distant, are Java and Sumatra, separated from Krakatau by a stretch of water about 25 miles in breadth. It is reasonable to wonder whether, had Forbes known of this and similar modern instances of the capabilities of plants as travellers, he would have adopted the view he did. In this connexion it may be added that in recent years the glaciation of Ireland has been shown to be more extensive than it was believed to be when Forbes wrote his essay.

There would seem to be no insuperable objection to the conclusion that the Mediterranean plants in Ireland and in the south of England reached their present home after the retreat of the ice at the end of the Glacial period, and after Ireland became an island. A full consideration of the problem is beyond the scope of this book, but I have briefly stated the case, not with the authority of an expert but in order to draw attention to a particularly fascinating study in plant-migration.

In a volume by W. Canton entitled *A Child's book of Saints* a story is told in which the presence in Ireland of Mediterranean species receives a more picturesque explanation. The Monk Bresal was sent to teach the brethren in a Spanish monastery the

Fig. 2. *Eriocaulon septangulare* With. West Connemara.
(Photograph by Mr W. R. Welch.)

music of Irish choirs. In later years Bresal longed
for a sight of his native land, though he loved his
home and 'every rock, tree, and flower' in his adopted
country. After returning to Ireland, his thoughts
reverted to Spain ; 'it appeared to him as though
he was once again in a granite nook among the rocks
beside the Priory'; he saw the ice-plant with its
little stars of white flowers sprinkled with red (the
London Pride) and a small evergreen tree from which
he had often gathered the orange-scarlet berries
(Arbutus). The Prior of the Spanish monastery 'with
heavenly vision saw Bresal gazing at the evergreen
tree and the ice-plant, and turning to the trees
blessed them and commended them to go and make
real his dream. As Bresal brushed away his tears
he saw with amazement at his feet the ice-plant and
hard by the evergreen tree.'

The plant represented in Fig. 2 is another
British species which tasks the ingenuity of students
of plant-geography. This is the Pipe Wort (*Eriocaulon septangulare*), the sole representative in
Europe of a certain family of Monocotyledons : it
flourishes in the west of Ireland and in the western
islands of Scotland but nowhere else in Europe ; it is
native on the other side of the Atlantic in Canada
and the northern United States of America. Mr
Praeger in describing the striking mixture of species
in the west of Ireland writes, 'The pool from which

we gather the American Pipe Wort is fringed with Pyrenean Heathers. The cracks which are filled with the delicate green foam of the maiden hair are set in Bearberry and Spring Gentian ; *Habenaria intacta,* far from its Mediterranean home, sends up its flower-spikes through carpets of mountain Avens ; and St Dabeoc's Heath and the dwarf Juniper straggle together over the rocky knolls '(25).

The presence of Eriocaulon on the western edge of Europe may be attributed to migration in pre-Glacial days from North America by way of a land-connexion, of which Greenland and Iceland represent surviving portions. The opinion held by Forbes, and advocated by some later naturalists, that the southern companions of Eriocaulon in the west of Ireland are survivors from a Tertiary flora which have lived through the Ice Age, is consistently extended to the Pipe Wort. On the other hand, before yielding to the temptation to regard these American and Mediterranean species as links with the Tertiary period, we must be convinced that the possibilities of post-Glacial introduction, even without the aid of land-bridges, have been exhausted. The Pipe Wort is a botanical puzzle which affords a good example of the accession of interest to field-botany effected by a knowledge of the distribution of the component members of the British flora. The problem of its past history suggests an experimental enquiry into

the adaptability of its seeds to dispersal, and em-
phasises the importance of the co-operation of
botanists and geologists in a common endeavour to
trace the origin of British plants.

In addition to the Pipe Wort, mention may be
made of three other American flowering plants
recognised in the Irish flora.   *Sisyrinchium angusti-
folium* recorded from the west of Ireland is native
in temperate North America ; the orchid, *Spiranthes
romanzoffiana,* met with in the south and north of
Ireland, is widely spread in Canada and the northern
States, while *Sisyrinchium californicum,* a native
of California and Oregon, was discovered by Mr Marshall in marshy meadow-land near Wexford (26).   In
the case of the more recently discovered American
immigrants, the possibility of human introduction
must be borne in mind, though there are no special
reasons for doubting that some, as in the case of Eriocaulon, reached the Irish coast by natural agencies.

CHAPTER III

THE GEOLOGICAL RECORD

' All the Epochs of the Past are only a few of the front carriages, and probably the least wonderful, in the van of an interminable procession.' J. B. BURY (*The Science of History*).

The portion of the earth's surface accessible to investigation is made up in part of accumulations of old sediments, some indistinguishable from the shingle, sand, and mud now in process of formation by the ceaseless action of denudation ; others have been hardened, gently folded or violently contorted and so far altered by crust-movements as to render their sedimentary origin well nigh unrecognisable. It is these sediments of former ages, the dust of lost continents, in which are preserved the majority of the fragmentary remains of plants and animals, the flotsam and jetsam of successive phases of evolution.

The crust of the earth, as Darwin wrote, 'with its imbedded remains must not be looked at as a well filled museum, but as a poor collection made at hazard and at rare intervals'[19]. It is from this imperfect record that we seek to discover the relative antiquity of the several groups or genera of living plants, and in the structure of extinct types we endeavour to discover connecting links between divisions of the plant kingdom which in the course of evolution have retained little or no signs of a common descent.

Sir Joseph Hooker in a letter to Darwin in 1859 speaks of his 'conviction that we have not in a fossilised condition a fraction of the plants that have existed, and that not a fraction of those we have are recognisable specifically'(12). Considering the nature of the palaeontological documents the wonder is how much they have taught us, and we may look with confidence to the results of future research in a field of which the importance has only recently been appreciated. With the strata of sedimentary origin are frequently associated igneous rocks, and in many continental regions, as in the majority of oceanic islands, the crust of the earth consists wholly of volcanic material or of rocks produced by the gradual solidification of molten magmas. Rocks composed mainly of carbonate of lime, such as limestones and chalk, bear witness to ocean beds or to sediments deposited on the floors of inland seas beyond the reach of land detritus where coral reefs were reared or the shells and other calcareous skeletons of animals supplied the material for future land. In such rocks the remains of calcareous seaweeds are frequently recognisable and occasionally, as in the English chalk, fragments of wood testify to transport from a distant land.

While there is little difficulty in explaining the nature of much of the earth's crust, in several parts of the world the strata are totally unfossiliferous and closely simulate crystalline rocks. In many cases it is believed that such strata represent ancient sediments which in the course of ages have been reduced by metamorphic agencies to a condition which has obscured or entirely obliterated all traces of their pristine state.

Since the pioneer work of William Smith, who in the early days of the nineteenth century first realised the importance of fossils as aids to the determination of relative age, geologists have devoted themselves to the task of correlating the sedimentary rocks of the world, using as criteria the order of superposition of the strata and the nature of their organic remains. The result has been to classify portions of the earth's crust into periods or chapters, which together constitute a record of geological evolution as complete as it is possible to obtain from the available data. The accompanying table shows the order of sequence of the epochs, which stand for terms of years of a magnitude beyond our powers to grasp.

The division of geological history into larger and smaller periods does not imply the recurrence of sudden revolutions ; it is in some measure dictated by considerations of convenience, but more particularly by our ignorance of certain stages in the history of the world due to the imperfection of the record.

GEOLOGICAL TABLE.

Showing the position in the Geological Series of the strata referred to in this volume.

TERTIARY (CAINOZOIC)	Recent Pleistocene	Superficial accumulations containing human remains (Metal age, Neolithic and Palaeolithic ages, Glacial deposits)
	Pliocene	Cromer Forest-bed, etc.
	Miocene	Absent from Britain.
	Oligocene	Bovey Tracey beds, etc.
	Eocene	London Clay, etc.
SECONDARY (MESOZOIC)	Cretaceous	Chalk Wealden beds
	Jurassic	Purbeck and Portland beds (Upper Jurassic) Oolites (Middle Jurassic) Lias (Lower Jurassic)
	Triassic	Rhaetic beds Keuper ,, (Marls with rock-salt, etc.) Bunter ,,

GEOLOGICAL TABLE (continued).

PRIMARY (PALÆOZOIC)	Permian	Red Sandstones, etc. Magnesian limestone
	Carboniferous	Coal Measures Millstone Grit Carboniferous limestone
	Devonian	Devonian limestones, etc. Old Red Sandstones
	Silurian	Sandstones, shales, some limestone
	Ordovician	Slates, sandstones, Volcanic rocks, etc.
	Cambrian	Slates, sandstones, etc.
	Pre-Cambrian or Archaean	Slates, Volcanic rocks, etc.

In certain parts of the world, as for example the north-west Highlands of Scotland, the Malvern Hills, Scandinavia, and in many other regions in Europe and North America, geologists have recognised what they believe to be the foundation stones of the world. These Archaean rocks, which underlie the oldest fossiliferous strata, belong to a period of geological evolution from which it appears to be hopeless to obtain any light as to the nature of the contemporary organic world. The earliest vestiges of life so far discovered exhibit a high degree of organisation, which unmistakably points to their being links in a chain extending far beyond the limits of the oldest

Cambrian strata in which recognisable fossils first occur. The rocks of the Cambrian and Ordovician epochs, as represented by the grits, shales, slates and other sedimentary strata in Wales, Shropshire, the Lake district and elsewhere, though in places rich in the remains of animals, afford no information in regard to the land vegetation. From the succeeding Silurian epoch very little evidence has been gleaned as to the nature of the flora, and it is not until we come to the sedimentary rocks of the Devonian era that records of plant-life occur in any abundance. The almost complete lack of botanical data in the pre-Devonian formations is in part due to the fact that these older rocks consist to a large extent of marine deposits formed under conditions unfavourable to the preservation of plants. That the land-surfaces of the older Palaeozoic eras supported an abundant vegetation there can be little doubt. The relics of plant-life furnished by the Devonian and succeeding formations represent the upper branching-systems of a deeply rooted and spreading tree, the lowest portions of which have been destroyed or have left no sign of their existence.

In descending the Geological series, we begin with superficial deposits, such as peat and river-gravels found subsequently to the underlying boulder-clay of the Glacial period. The remains of forest trees preserved in the peat and in submerged forests round

the coast connect the vegetation of the historic period
with that of the Neolithic age.   At the base of the
Pleistocene series, the name given to the latest
chapter of geological history, we find evidence of the
prevalence of arctic conditions in the widely spread
boulder-clays and other deposits of the Glacial period.

From deposits of post-Glacial date abundant plant
remains have been obtained, but we cannot say with
any degree of certainty what proportion of these
plants remained in Britain during the Ice age, and
whether the greater part of the vegetation, the relics
of which have been discovered in pre-Glacial beds,
was destroyed or driven south by the advancing ice.
We may briefly consider some of the more interesting
facts brought to light by the investigation of the
fossil plants in the Lower Pleistocene and Upper
Tertiary beds.   It is mainly to the researches of
Mr Clement Reid into the vegetation of Britain
immediately preceding the Glacial period, that our
knowledge of this phase of the history of the British
flora is due.

On the coast of Norfolk in the neighbourhood of
Cromer the sections of the cliffs reveal the existence
of a succession of sands, clays, and gravels underlying
Glacial deposits ; this material was probably laid
down near the mouth of the ancient Rhine, which in
the latter part of the Tertiary period flowed across
a low area, which is now occupied by the shallow

southern half of the North Sea(27). The plant-
fragments found in these river-sediments indicate a

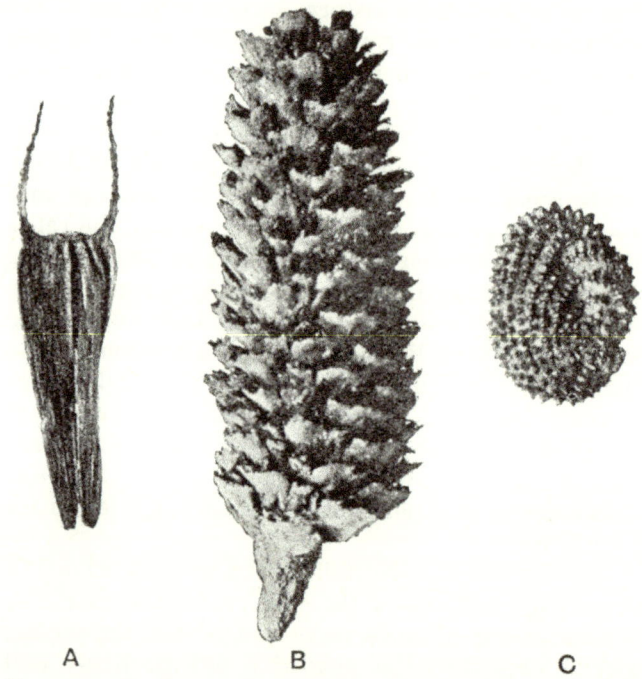

Fig. 3. Pre-Glacial plants from Mundesley (A), Norfolk and Pakefield
(B, C), Suffolk. (Photographs by Mr Clement Reid and Mrs
E. M. Reid.) A. *Bidens tripartita* Linn. (×6); B. *Picea excelsa*
Linn. (nat. size). C. *Stellaria holostea* Linn. (×12).

temperate climate. Among the plants of this pre-Glacial flora are many familiar British species, such as *Caltha palustris* (marsh marigold), species of buttercup, *Stellaria holostea* (greater stichwort) (Fig. 3, C), *Bidens tripartita* (bur-marigold) (Fig. 3, A), maple,

Fig. 4. *Trapa natans* Linn. (nat. size). From Mundesley.
(Photographs by Mr and Mrs Reid.)

hawthorn, the alder, hazel, the yew, Scots pine and numerous others. If, as is not improbable, these pre-Glacial plants were swept away by the subsequent arctic conditions, the great majority of them returned to their old home when a warmer climate ensued.

There are however some pre-Glacial plants, such as the spruce fir (*Picea excelsa*), a cone of which is shown in Fig. 3, B, the water chestnut, *Trapa natans* (Fig. 4), and a few other species no longer represented in the British flora. The genus *Trapa* is a striking example of a flowering plant which has disappeared since the Tertiary period from many parts of Europe, including England, most of Sweden, and from several regions in northern Europe. It still grows in a few localities in Switzerland and in some of the Italian lakes. In pre-Glacial times the water chestnut was widely spread from Portugal and England in the west to Siberia in the east, and its hard four-pronged nuts have been recorded from many post-Glacial peat-moors in the north of Europe.

From the so-called Cromer forest-bed and associated deposits on the Norfolk coast several pre-Glacial plants have been obtained, indicating a temperate climate during this phase of the Pleistocene period. A few arctic species, such as the dwarf birch and arctic willow obtained from the deposits next above the Cromer forest-bed, herald the near approach of glacial conditions.

It may be remarked in passing that no satisfactory evidence has been discovered in Britain of the existence of man in this part of Europe in pre-Glacial days: it is, however, believed that flints from Tertiary strata on the continent show traces of human workmanship. As Sir Edwin Ray Lankester said in 1905, 'It is not improbable that it was in the remote period known as the Lower Miocene—remote as compared with the gravels in which Eoliths [primitive stone implements] occur—that Natural Selection began to favour that increase in the size of the brain of a large and not very powerful semi-erect ape' (28).

Though comparatively recent in terms of geological chronology, the remoteness, according to ordinary conceptions of time, of the Tertiary period is brought home to us when we endeavour to grasp the fact that it was during this chapter in the earth's history that some of our highest mountain-ranges, such as the Alps, the Carpathians, and Himalayas were formed by the uplifting of piles of marine sediments. From Tertiary strata in the Isle of Wight, on the Hampshire coast, and in the London basin numerous fossil plants have been obtained, which afford convincing evidence of climatic conditions much more genial than those of the present day. The presence of palm leaves and of a wealth of other sub-tropical plants in Lower Tertiary beds in England reveals the existence of a flora differing considerably both from that in the uppermost Tertiary beds of Norfolk and from the modern British flora, but closely allied to the present Mediterranean flora.

The basaltic columns of the Giants' Causeway and of the Staffa Cave, and the terraced rocks which form so characteristic a feature in the contours of the Inner Hebrides, are portions of lava-flows, which in the early days of the Tertiary period were poured out over a wide area of land stretching from the north-east of Ireland, through the western isles of Scotland, the Faroë islands, to Iceland and Greenland. While in this northern region volcanic activity was being manifested on a stupendous scale, a shallow sea extended over part of what is now the south-east of England in which was deposited a considerable thickness of sedimentary material derived from the neighbouring land. In this upraised sea-floor, known as the London clay, which is exposed in the Isle of Sheppey and in many other localities, numerous fossil fruits and fragments of wood occur in association with marine shells. The fact that many of the fruits were ripe at the time of their entombment led some eighteenth century writers to assign an autumn date to the universal deluge. One of the Sheppey fruits may be mentioned as an especially interesting sample of the early Tertiary flora, namely the genus Nipa-dites, so named from the very close resemblance of the fossils to the fruits of the existing tropical plant Nipa. *Nipa fruticans*, sometimes described as a stemless palm because of the absence of the erect stem which is usually a characteristic feature of palms, grows in brackish estuaries of many tropical countries (Fig. 5, A): it has long leaves not unlike

A   B

Fig. 5. *Nipa fruticans*, Thunb. A. On the coast of the Malay Peninsula. (Photograph by Prof. Yapp.) B. Head of fruits (⅓ nat. size). From a specimen in the British Museum.

4—2

those of the date-palm and bears clusters of fruits
as large as a man's head; a single fruit is two or
three inches long and its hard fibrous shell is characterised by four or five longitudinal ribs (Fig. 5, B).
The fruits of Nipa, which may be carried a considerable distance by ocean-currents without losing the
power of germination, are constantly found with other
vegetable drift on the beaches of tropical islands. The
discovery of fruits of Nipa (or Nipadites), hardly
distinguishable from those of the existing species,
in Tertiary beds in England, Belgium, in the Paris
basin, and in Egypt affords a striking instance of
changes in the geographical distribution of an ancient
plant now restricted to warmer regions.

While the higher members of the Cretaceous
system, as seen in the chalk cliffs and downs, represent the upraised calcareous accumulations on
the floor of a fairly deep and clear sea, the lower
members testify to shallower water within reach of
river-borne sand and mud. 'During the Chalk
period,' as Huxley wrote, 'not one of the present
great physical features of the globe was in existence.
Our great mountain ranges, Pyrenees, Alps, Himalayas, Andes, have all been upheaved since the
chalk was deposited, and the Cretaceous sea flowed
over the sites of Sinai and Ararat' (29).

The Wealden strata, at the base of the Cretaceous
system, as seen on the Sussex coast, in parts of the

Isle of Wight, in the Weald district of Kent and neighbouring counties, point to the existence of a lake over a portion of the south of England and of the English Channel. The remains of a rich Wealden flora have been collected from these Wealden sediments, notably from the plant-beds of Ecclesbourne near Hastings, in which, so far as we know, flowering plants played no part or at most occupied a very subordinate position. A few fossil leaves have been described from rocks assigned to a Wealden age, —and from the older Stonesfield Slate, of Jurassic age, a single leaf is recorded,—which seem to be those of Dicotyledons ; but it is certain that even in the early days of the Cretaceous period the present dominant group in the plant kingdom was in its infancy and in many regions probably unrepresented. When we glance at the geological table and consider that in all the floras from the Wealden down to the Devonian period, flowering plants played no part, we are able to appreciate the fact of their rapid development, referred to in a previous chapter, when once this highest type had become established.

The rocks comprised in the Jurassic system extend from East Yorkshire to the coast of Dorsetshire ; they consist of a succession of limestones, clays, sandstones, and a few thin beds of impure coal. Sediments of this age also occur, though to a much less extent, on the north-east coast of Scotland and in a few places in the Inner Hebrides. Many of the Jurassic strata contain only marine shells, and corals are occasionally abundant, though in the lower members of the system in the cliffs near Lyme Regis and at Whitby fossil plants are fairly common. It is, however, from the middle Jurassic beds, in the cliffs between Whitby and Scarborough, and in some inland quarries in East Yorkshire, that we have obtained the richest Jurassic flora. Rivers from a northern land laden with sediment and carrying driftwood, leaves and other plant fragments, deposited their burden in an estuary which occupied the eastern edge of Yorkshire. Sedimentary rocks laid down towards the close of the Jurassic period in the island of Portland in the south and on the Sutherland coast in the north have furnished valuable records of plant-life.

The passage from the Jurassic to the underlying Triassic system is formed by some shales and limestones in South Wales containing remains of fish and other marine organisms. These so-called Rhaetic beds are poorly represented in the British area, but on the continent of Europe and in other regions the sediments of this age bulk much more largely and have yielded a rich collection of plants. The rocks of the upper division of the Triassic system, as seen in the Midlands, point to the prevalence of desert conditions; and in the grooved sand-polished surfaces of granite in Charnwood forest we have a glimpse of a Triassic landscape. The salt-bearing strata of this period in Cheshire and Worcestershire suggest conditions paralleled at the present day in the Caspian and Dead-Sea regions. The vegetation of Britain, and indeed of the world as a whole, seems to have undergone but little change during the enormous lapse of time represented by the sediments comprised between the Wealden and Triassic periods. The Lower Triassic flora affords evidence of a change in the facies of the vegetation and prepares us for the still greater differences revealed by a study of the Permian and Carboniferous floras. To the student of evolution these Palaeozoic floras are of special interest on account of the facts they have contributed in regard to the descent and inter-relationship of different branches of the vegetable kingdom.

It is by a patient study of the waifs and strays of the vegetation of successive phases of the world's history preserved in sedimentary strata, that it has been possible to follow the history of many existing plants and to establish links between the present and the past.

CHAPTER IV

PRESERVATION OF PLANTS AS FOSSILS

'Some whim of Nature locked them fast in stone for us after-thoughts of Creation.' LOWELL.

The failure of the earlier naturalists to grasp the true significance of fossils or even to appreciate their nature is an extraordinary fact when we consider the pioneer work which they accomplished in biological and geological science. The following extract from the writings of so enlightened a man as John Ray serves to illustrate an almost incredible disinclination to admit what seems to us the obvious. He wrote :—
'Yet I must not dissemble that there is a Pheno-menon in Nature, which doth somewhat puzzle us to reconcile with the prudence observable in all its work, and seems strongly to prove, that Nature doth sometimes *ludere*, and delineates figures, for no other end, but for the ornament of some stone, and to entertain or gratify our curiosity, and exercise our wits. This is, those elegant impressions of leaves and

plants upon cole-slate, the knowledge of which,
I must confess myself to leave to my learned and
ingenious friend Mr Edward Lhwyd of Oxford....He
told me that Mr Woodward, a Londoner, shewed
him very good draughts of the common female fern,
naturally formed in cole....But these figures are more
diligently to be observed and considered...Dr Woodward will have them to be the impressions of the
leaves of plants which were there lodged at the time
of the Deluge '(31).

The Mr Woodward alluded to by Ray thus
expressed his views on fossils in an *Essay towards
the Natural History of the Earth* :—'The whole
terrestrial globe was taken all to pieces and dissolved
at the Deluge, the particles of stone, marble, and
all solid fossils dissevered, taken up into the water,
and then sustained together with sea shells and other
animal and vegetable bodies ; the present earth
consists and was formed out of that promiscuous mass
of sand, earth, shells, and the rest falling down again,
and subsiding from the water ' (32).

In the later part of the seventeenth century
Steno, a Dane by birth and Professor of Anatomy at
Padua, by his recognition of the identity of the teeth
in a shark's head, which he had dissected, with some
fossils from Malta known as Glossopetrae, established
the true nature of fossils.   He also recognised a
certain orderly sequence in fossiliferous strata, and

in the opinion of Professor Sollas he is entitled to
be considered the 'Father and Founder' of Geology(33).

It was by slow degrees that the early observers
freed themselves from the obsession that the remains
of animals and plants in the earth's crust bear witness
to a Universal Deluge and are all identical with
existing species. The possibility that some of the
fossil plants in English strata might be more clearly
related to forms now met with in warmer regions
was gradually realised. The publication of the
Origin of Species stimulated palaeontological
research, and botanists as well as zoologists turned
to the investigation of extinct genera in search of
proofs of the doctrine of evolution.

The common occurrence of petrified wood in
rocks of different ages is well known. Fossil stems
are occasionally found in their natural position of
growth, the structural details being rendered permanent by the deposition of siliceous or calcareous
material from water drawn by capillarity into the
dead but still sound tissues. Petrified wood from
Upper Jurassic beds is abundant in the Island of
Purbeck ; an unusually long piece of stem may be
seen in the small town of Portland fixed to the wall
of a house. Some of these stems have been referred
by an American author to the Araucarian family of
Conifers, but the structure is as a rule hardly well
enough preserved to afford satisfactory evidence for

identification.   In his *Testimony of the Rocks,* Hugh
Miller speaks of fossil wood from the upper beds of
the Jurassic system in sufficient abundance on the
beach at Helmsdale in Sutherlandshire to be collected
in cart-loads ; it is still easy to pick up good specimens on the shingle beach a short distance north of
Helmsdale, and a recent microscopical examination
showed that some specimens are pieces of an Araucarian tree.

Impressive examples of petrified trees on a
large scale are to be seen in the United States,
in Arizona and the Yellowstone Park.   (Frontispiece.)
In the northern part of Arizona the country for over
an area of 10 square miles is covered with tree trunks,
some reaching 200 feet in length and a diameter of
10 feet. The nature of the mineralising substance has
given rise to the name Chalcedony Park for this
Triassic forest (34).   A striking example of one of the
Arizona trees is exhibited in the British Museum and
in a neighbouring case is a splendid petrified stem, 9 ft.
in height, of a conifer discovered in Tertiary lavas in
Tasmania (35).

Figure 6 illustrates the preservation of a series
of forests of Tertiary age in the mass of volcanic
sediments, 2000 feet in thickness, known as Amethyst
mountain, in the Yellowstone Park district.   By the
weathering away of the surrounding volcanic material
the tall stems of the trees are exposed in places on

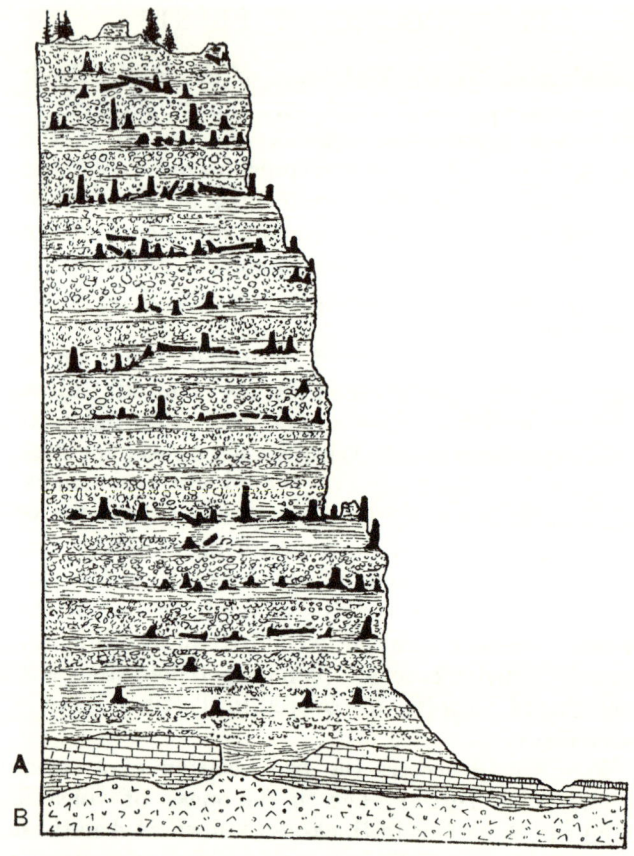

A

B

Fig. 6. Section of the north face of Amethyst Mountain, Yellowstone Park, including upwards of 2000 ft. of strata. The
steepness of the slope is exaggerated. (After W. H. Holmes.)

the mountain sides like the 'columns of a ruined temple.' The height of the river at the foot of the cliff is 6700 ft. above sea-level and the mountain rises to a height of 9400 ft. above the sea. In the lower part of the section the volcanic strata are seen to rest on a foundation of older rocks A, and these in turn were laid down on the eroded surface of a still more ancient foundation, B(36).

The section as a whole affords a striking demonstration of the magnitude of earth-movements since the last of these forests was buried below the surface of a sea in which the volcanic material was deposited. The account of the Yellowstone Park section recalls Darwin's description(37) of snow-white columns projecting from a bare slope, at an altitude of 7000 ft. in the Cordillera.

The abundance of drift-wood on the coasts of some countries at the present day helps us to picture the conditions under which the remains of former forests have been preserved. In his *Letters from High Latitudes*, Lord Dufferin gives the following description of drift-wood on the shores of Spitzbergen :—'A little to the northward, I observed, lying on the sea-shore innumerable logs of drift-wood. This wood is floated all the way from America[1] by the Gulf Stream, and as I walked from one hugh bole to another, I could not help wondering in what

[1] My friend, Prof. Nathorst, of Stockholm, tells me that the wood probably came from Siberia.

primeval forest each had grown, what chance had originally cast them on the waters, and piloted them to this desert shore'(38). A photograph reproduced in Amundsen's book on *The North West Passage* shows the beach on the Alaskan coast strewn with drifted timber(39). For the accompanying photograph (Fig. 7) of the flood-plain of the Colorado River(40), I am indebted to Professor MacDougal of the Desert Research Laboratory at Tucson, Arizona, who in a recent letter writes, 'During times of high-water a thin sheet of flood covers the flat for many miles and bears drift-wood so thickly that it is difficult to push a boat through it.' The drift-wood consists of poplar, willow, pine, and juniper, 'the last two have been brought from the upper river, from as far away as a thousand miles.' A picture such as this affords an admirable example of the wealth of material available for preservation in a fossil state.

It is only in the minority of cases that the accidents of preservation of fragments of ancient floras have given us the means of investigating the internal structure of the plant organs. It is far more frequently the case that fossil plants are represented only by a carbonised film on the surface of a piece of shale or other rock: the actual substance of the plant has been converted into a thin layer of coal, and though the venation and other surface-features may be clearly revealed, the internal tissues have

Fig. 7. Flood-plain of the delta of Rio Colorado. The hills in the background are 25 miles distant. (From a photograph by Prof. MacDougal.)

been destroyed. If a lump of clay containing a piece of fern frond is heated, the result is an impression of the leaf on the hardened matrix and a coaly substance in place of the plant substance. It is occasionally possible by detaching a piece of the black film from a fossil, and heating it with nitric acid and chlorate of potash and then dipping it in ammonia, to obtain a transparent preparation suitable for microscopical examination of the cell-outlines of the superficial layer of the leaf or other plant-fragment. This method of examination, used by several students of fossil plants and with conspicuous success by Professor Nathorst of Stockholm, often affords valuable aids to identification.

Pieces of plants embedded in sandy sediment, if not preserved by petrifaction, that is by the introduction into the tissues of some siliceous or calcareous solution, gradually decay and their fragmentary remains may be washed away by percolating water, leaving a hollow mould in the gradually hardening sediment, which is afterwards filled with sand or other material. The plant itself is destroyed, but a cast is taken which in the case of fine-grained sediments reproduces the form and surface-pattern of the original specimen. The incrustation of plants by the falsely named petrifying springs of Knaresborough and other places illustrate another method of fossilisation.

Plants which owe their preservation to amber occur both as incrustations and petrifactions. This fossil resin occurs in Tertiary, Cretaceous, and Jurassic rocks; the amber found in abundance on the Baltic coast near Danzig and occasionally picked up on the beach in Norfolk and Suffolk comes from

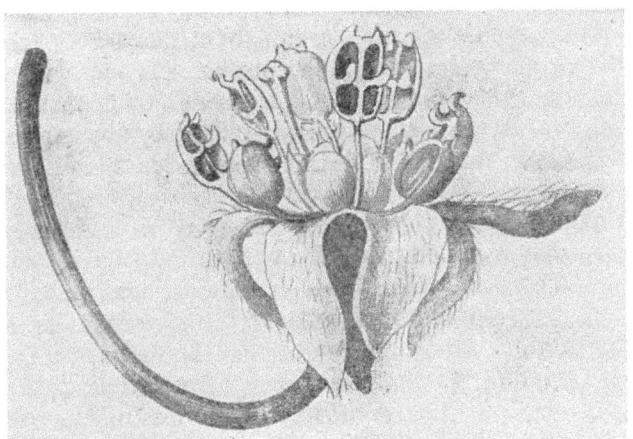

Fig. 8.   Flower of *Cinnamomum prototypum* Conw. preserved in amber.   × 10.   (After Conwentz.)

beds of Tertiary age. Pieces of Pine-wood have been described from the Baltic beds in which the tissues are perfectly preserved as the result of the conversion into amber of the resinous secretion which permeates their cells: in this case the amber is a petrifying

S.   5

agent.   More frequently the preservation is due to
incrustation ; as resin trickled down the stems of the
Tertiary pines from an open wound, flowers and
leaves, blown by the wind on to the sticky surface,
were eventually sealed up in a translucent case of
amber.   Though the actual substance may have gone,
the mould which remains exhibits in wonderful per-
fection each separate organ of a flower or the
delicate hair-clusters on the surface of a leaf.   The
flower represented in Fig. 8, a species of Cinnamon,
is one of several specimens described by the authors
of a monograph of Tertiary plants in the Baltic
amber[41].

The fragments of plants preserved in nodules of
calcareous rock occasionally met with in some of the
Lancashire and Yorkshire coal-seams are perhaps
the most striking examples of the possibilities of
petrifaction.   By cutting sections of these nodules
and grinding them to a transparent thinness, the
most delicate tissues of Carboniferous plants are
rendered accessible to investigation under the high
power of a microscope.   As our attention is absorbed
by the examination of the details of cell-structure
it is easy to forget that the section has not been
cut from a living plant, but from the twig of a
tree which grew in the forests of the Coal age.   The
preservation is such as to enable us not only to
describe the anatomy of these extinct types of

vegetation, but, by the application of the knowledge of the relation between the structure of the plant-machine and its functions gained by a study of living species, it is possible in some degree to picture the plants of the Coal period as living organisms and to see in the structural framework a reflection of external environment. The recognition in the general architectural plan of the Palaeozoic plants, as in many of the finer anatomical features, of the closest resemblance to plants of the modern world produces an almost overwhelming sense of continuity between the past and the present.

The plants of the Palaeozoic period, though often differing considerably from those of the same class in the floras of to-day, exhibit a remarkably high type of organisation. Some of the most abundant trees in the forest of the Coal age are decidedly superior in the complexity of their structure, as also in size, to modern survivals of the same stock. On the other hand, it must be remembered that Mono-cotyledons and Dicotyledons which now occupy the highest place in the hierarchy of plants have left no sign of their existence in any of the Palaeozoic strata. The greater size of some of the Palaeozoic plants, and in some respects the more advanced stage of evolution which they represent as compared with their nearest relatives of the present era, must be considered in relation to their more important and relatively higher position in the plant-world than that which is now held by their diminutive descendants. It is, however, impossible to get away from the conclusion that the oldest Palaeozoic flora of which we have an intimate knowledge must be the product of development of an age which is represented by a chapter in the history of the plant kingdom at least as far removed from the beginning as it is separated from the chapter now being written. Examples might be quoted in illustration of the risks attending the determination of fossils by means of external features alone, but it may suffice to mention the case of a specimen originally described as a fragment of a Cretaceous Dinosaur under the name *Aachenosaurus multidens*. By the examination of thin sections this supposed bone was shown to be a piece of Dicotyledonous wood(42). The methods of preservation of plants as fossils are numerous and varied and the few examples selected give but an incomplete idea of the subject : for a fuller treatment of fossilisation the reader is referred to more technical treatises (48).

The employment of fossil plants as 'Thermometers of the ages' is a branch of Palaeobotany to which a passing allusion may be permitted though it is only indirectly connected with the main question. As one of the most interesting examples of changed climatic conditions revealed by a study of fossil plants, reference may be made to the wealth of material collected within the Arctic circle. The problems suggested by the discovery of plants in rocks of various ages in North Siberia, Spitzbergen, Franz Josef Land, Bear Island, Greenland, and in many other localities in the far north are too difficult and far-reaching to be discussed in these pages. In the Cretaceous and Tertiary strata of the west coast of Greenland and Disco Island from 69° to 72° north latitude, to refer only to one case, a great number of plants have been obtained by several of the earlier Arctic explorers and more recently by members of one of the Peary Expeditions. At the present day on the fringe of land on the western edge of Greenland which is not permanently covered with ice, a considerable number of herbaceous plants are able to exist and to produce seed during their concentrated period of development; while trees are represented only by a few low-growing shrubs such as the dwarf Juniper. In places accessible to investigation beyond the ice-covered hills of northern Greenland the rocks have been shown to consist of Cretaceous and Tertiary sediments containing fossil plants associated with seams of coal. From these beds numerous Dicotyledons have been obtained, some of them almost identical with living species characteristic of sub-tropical or tropical countries. In the lowest of the Cretaceous series no Dicotyledons have been found, but flowering plants are abundant in the higher Cretaceous rocks. Allowing for the fact that closely allied species are often able to live under very different climatic conditions, there can be no doubt that the Cretaceous and Tertiary floras of Greenland indicate an average temperature considerably higher than that which now prevails in the warmest parts of the British Isles.

In the far south a fairly rich Jurassic flora has recently been discovered by the members of a Swedish Antarctic expedition in Graham's Land in latitude 63°·15 S. and longitude 57° W., which in its general facies bears a close resemblance to the Jurassic flora of Yorkshire.

Although the great majority of the records of ancient plants are difficult to interpret by reason of imperfect preservation and because of the frequent separation of leaves, stems, and reproductive organs, the student who tries to piece together the *disjecta membra* of the floras of the past shares the opinion expressed by the late Marquis of Saporta,—'Si l'on s'attache à les déchiffrer, on oublie bien vite la singularité des caractères, et le mauvais état des pages. La pensée se lève, les ideés se développent, le manuscrit se déroule ; c'est la tombe qui parle et livre son secret.'

CHAPTER V

'It has been shown that certain forms persist with very little change, from the oldest to the newest fossiliferous formations; and thus show that progressive development is a contingent, and not a necessary, result of the nature of living matter.' HUXLEY.

The Ferns as a whole represent a section of the vegetable kingdom which traces its ancestry as far into the past as any group of plants. Impressions of leaves on the shales of the Coal-measures and on rocks of the earlier Devonian period are hardly distinguishable in form and in the venation and shape of the leaflets from the finely divided fronds of modern ferns. Until a few years ago these Palaeozoic fossils were generally regarded as true ferns, and it was believed that ferns played a conspicuous part in the vegetation of the earliest periods of which we have any botanical knowledge. Conclusions based on external form must frequently be revised in the light of more trustworthy evidence. It was shown in the later part of the nineteenth century by the late Professor Williamson of Manchester, whose researches into the plants of the Coal age shed a flood of light on the ancestry and inter-relationship of many existing plants, that some of the fern-like leaves which have long been familiar to those who search among the shales of the refuse heaps of collieries, were borne on stems differing in anatomical features from those of any known fern. The investigation of the structure of the leaves and their supporting stems led to the recognition of certain extinct genera of Palaeozoic plants of exceptional interest, to which the term generalised type is aptly applied. Associated with anatomical and other characters such as we now regard as the attributes of ferns, these plants exhibit other features not met with in modern ferns but characteristic of a group of seed-bearing plants known as the Cycads. Recent research has revealed the existence of several such generalised types which, by their combination of characters now met with in distinct sub-divisions of the plant-kingdom, clearly indicate the derivation of Ferns, and Cycads as we know them to-day, from a common stock. It was in the first instance by means of anatomical evidence—obtained by the microscopical examination of sections of petrified fragments of stems and leaves—that the generalised nature of these Palaeozoic plants was recognised. Nothing was known as to the reproductive organs.

Ferns as now represented in the floras of the world are essentially seed-less plants. As the author of *Hudibras* wrote :

> 'Who would believe what strange bugbears
> Mankind creates itself, of fears?
> That spring like fern, that insect weed,
> Equivocally, without seed.'

The reproductive organs or spores borne on the fronds of a fern produce, on germination, a thin green structure, known as the prothallus, less than an inch in length : this bears the sexual organs, and as the result of the union of the male and female cells, the embryo fern-plant begins its existence as a parasite on the inconspicuous prothallus, until after unfolding its first green leaf and thrusting a slender root into the ground, it starts its career as an independent organism[1]. In this life-cycle the seed plays no part.

It is noteworthy that the absence of any indication of spore-capsules and spores, in the case of some of the supposed fern leaves from the Coal-measures, caused some suspicion in the mind of an Austrian Palaeobotanist as to the right of such specimens to

---

[1] The life-history of a Fern is clearly described by Prof. Bower in a recent volume in this series.

be classed among the ferns. This opinion, based in
the first place on negative evidence and but little
regarded by other authors, has in recent years
been proved correct. In 1904 a paper was read
before the Royal Society by Professor Oliver and
Dr Scott[43] in which evidence was brought forward
pointing to the conclusion that one of these
generalised plants bore true seeds. Subsequently
Dr Kidston published an account of some specimens
of another of these Palaeozoic plants in which was
actually shown an organic connexion between undoubted seeds and pieces of a fern-like frond[44].
Without entering into further details, these and
similar discoveries may be summarised as follows :—

Many of the supposed Fern-fronds of Palaeozoic
age, particularly those characteristic of the Coal-
measures, are the leaves of plants which in their
anatomical characters combined features now shared
by true Ferns and by the Cycads. The reproductive
organs of these Palaeozoic genera differed widely
from those of existing ferns ; the male organs, while
not unlike the spore-capsules and spores of certain
ferns, recall the male organs of living Conifers and
Cycads, and the female organs were represented by
seeds of a highly complex form. These seed-bearing
plants have been called Pteridosperms, a name
which expresses the combination of fern-like features
with one of the distinguishing attributes of the

higher plants, namely the possession of seeds. The ancestors of Pteridosperms are as yet unknown; it is, however, reasonable to assume that there existed in some pre-Carboniferous epoch a group of simple plants from which both Ferns and Pteridosperms were derived. In the forests of the Coal age true Ferns probably occupied a subordinate position in relation to the Pteridosperms.

The question of the relationship between different families of recent ferns and the older known fossil members of the group is beyond the scope of this book. Evidence has been discovered in recent years which warrants the statement that, although none of those Carboniferous ferns were generically identical with existing forms, they very clearly foreshadowed some of those structural features which characterise more than one family of present-day Ferns. The records of the older Mesozoic formations afford abundant evidence of the existence of certain types of Ferns showing a very close resemblance to recent species.

An enquiry into the geographical distribution of living Ferns reveals facts of special interest in connexion with the relative antiquity of different genera and families. The wide distribution of the Bracken fern has already been referred to: it is abundant in Tasmania; its vigour in the island is well illustrated by Mr Geoffrey Smith's statement

Fig. 9. *Osmunda regalis* Linn. Fertile frond. (⅔ nat. size.)

that constant attention is necessary to keep it from invading newly opened country(45). On Mount Ophir in the Malay Peninsula the cosmopolitan bracken occurs in association with the two genera Matonia and Dipteris, ferns which are among the most striking examples of links with a remote past and have a restricted geographical range. With *Osmunda regalis*, the Royal Fern, the Bracken is conspicuous in the marsh vegetation of the Bermudas; it flourishes on the Atlas Mountains, in the Canary Islands, in Abyssinia, on Mt Kenia, in British East Africa, in the Himalayas, in Persia and China, in New Zealand, and is in fact generally distributed in the tropics as also in both the north and south temperate zones.

The Royal Fern (Fig. 9) is another British species with a wide distribution; it occurs in Northern Asia and in North America; it is common in the Siberian forests and lives in several tropical countries, extending to Southern India and Cape Colony, and in South America it is represented by a closely allied species. Though at the present day *Osmunda regalis* is one of the rare English Ferns, its occurrence in the submerged forest-beds round our coasts and in pre-Glacial beds points to its former abundance in the British area generally. The Royal Fern is a member of a family now represented by two genera, Osmunda and Todea.

With the exception of *Todea barbara*, with its

large spreading fronds and a short root-covered stem, which occurs in Australia and Cape Colony, all the species of this genus are filmy ferns with semi-transparent fronds, like those of the British fern *Hymenophyllum tunbridgense*, adapted to a moisture-laden atmosphere. The maximum development of the genus is in New Zealand.

Todea barbara affords an instance of discontinuous distribution ; it was no doubt once widely spread in circumpolar regions and now survives only in South Africa and in Australia.

There are satisfactory reasons for regarding the Bracken Fern, with its world-wide range in present-day floras, as a comparatively modern species now in full vigour. Its anatomical and other features are consistent with the view that it is a late product of evolution, and as yet no indication has been given by the records of the rocks of an ancient lineage. The Osmunda family, on the other hand, is undoubtedly an extremely old branch of the fern group. A comparison of the Royal Fern with the Bracken shows that their stems are constructed on very different plans, and we have good reasons for speaking of the structural peculiarities of the former as those of a more primitive type. Moreover, the discontinuous geographical range of some members of the Osmunda family is in itself an indication of antiquity. There

is another point which may have a bearing in this
question of antiquity, namely the fact that the spores
of Osmunda are green and do not possess the powers
of endurance inherent in the spores of the majority of
ferns which are not green.   It has recently been con-
tended by Professor Campbell of Stanford University
that the delicate green spores of the Liverworts,
plants closely allied to the Mosses, constitute an
argument in favour of the antiquity of these plants[46].
Certain Liverworts are cosmopolitan in their range,
e.g. the genera Riccia and Marchantia.

If certain genera are widely distributed, notwith-
standing the fact that their reproductive cells, by
which dispersal is effected, are ill-adapted to withstand
unfavourable conditions or to endure prolonged
desiccation, it would seem reasonable to conclude
that their emigration has been accomplished slowly
and with difficulty.   Ferns such as Osmunda, with
green and short-lived spores, would thus be handi-
capped in competition with other genera provided
with more efficient means of dispersal and better
equipped for the vicissitudes of travel.

The inferences as to antiquity deduced from a
study of the existing species of Osmunda and Todea
receive striking confirmation from the testimony of
fossils.   Some of the oldest known Palaeozoic ferns,
though differing too widely from the existing Osmundas
and Todeas to be included in the same family, afford

distinct glimmerings of Osmundaceous characters, which at a later period became individualised in the direct ancestors of the modern forms. Our knowledge of the past history of the Osmunda family has recently been considerably extended and placed on a firmer basis by the researches of Dr Kidston and Professor Gwynne-Vaughan. These authors have recognised in some exceptionally well-preserved fern-stems from Permian rocks in Russia, anatomical features which point unmistakably to close relationship with the recent members of the family [47][48].

Passing higher up the geological series, fertile fern fronds with spore-capsules and spores practically identical with those of Osmunda have been found in the Jurassic plant-beds of Yorkshire and in rocks of approximately the same age in many parts of the world. From Jurassic strata in New Zealand a petrified fern-stem has been described (*Osmundites Dunlopi*), almost identical in structure with the surviving species. Cretaceous and Tertiary examples of similar ferns might be quoted; but enough has been said to establish the claim of the Royal Fern and other members of the Osmunda-family to an ancestry which possibly extends even farther back than that of any other existing family of Ferns.

A brief reference may be made to another fern now represented by several species widely disseminated in tropical and sub-tropical countries. The genus

Gleichenia occurs abundantly in the warmer regions
of both the Old and New World. The fronds may
usually be recognised by their habit of growth
(Fig. 10); in several species the main axis is

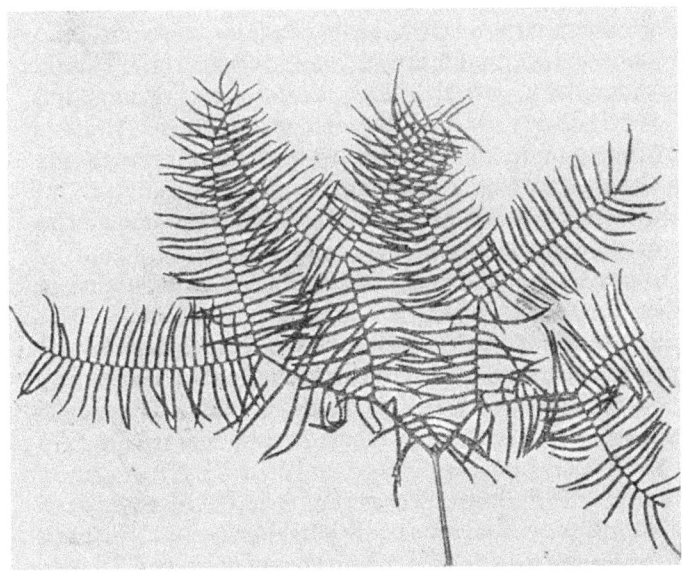

Fig. 10. *Gleichenia dicarpa* Br. (½ nat. size.)

repeatedly forked and a small bud between the
divergent branches of the forks forms a characteristic
feature. The leaflets are either long and narrow

s.   C

like the teeth of a comb or short and bluntly rounded. Moreover the anatomy of the creeping stem affords a ready means of identification. We have satisfactory evidence of the occurrence of Gleichenia in European floras during both the Jurassic and Cretaceous periods. Numerous fragments of plants were obtained some years ago, not far from Brussels, from the Wealden strata in which the famous skeletons of Iguanodon were discovered. Visitors to the Natural History Museum in Brussels are no doubt familiar with the skeletons of this enormous herbivorous animal : in the same gallery are exhibited the remains of the fossil plants from the Iguanodon beds. Some of these fragments are pieces of fern fronds identical in form with those of existing Gleichenias. The microscopical examination of some exceptionally well preserved fragments of Wealden stems discovered by Prof. Bommer of Brussels enabled him to recognise the Gleichenia type of structure and thus to confirm the inconclusive evidence furnished by fragmentary leaves. The most interesting records in regard to the former occurrence of Gleichenia in Northern Europe we owe to the late Oswald Heer of Zürich, who has described many examples of Gleichenia fronds from rocks of Lower Cretaceous age in Disco Island on the west coast of Greenland in latitude 70° N. The same type of fern is recorded also from upper Jurassic beds in the north-east of

Scotland, in the Wealden rocks of Sussex, as well as
from other European localities. It is clear that the
Gleichenia-family, no longer represented in north
temperate floras, was in the Jurassic period, and
especially in the early days of the Cretaceous period,
widely spread in Europe, extending well within the
Arctic circle. It may be that the original home of
Gleichenia was in the far North at a time when
climatic conditions were very different from those
which now prevail. Gleichenia, like many other
northern plants, retreated to more southern regions
where, in the warmer countries of the world, many
species still flourish widely separated in space and
time from the place of their birth.

The ferns so far mentioned have a more or less
extended distribution at the present day. In the
case of *Pteridium aquilinum*, the cosmopolitan
Bracken Fern, wide range would seem to be corre-
lated with comparatively recent origin; on the
other hand, the facts of palaeobotany show that
the wide distribution of Osmunda, a type of fern
which differs in many important respects from
members of the family (Polypodiaceae) to which
the Bracken belongs, is not inconsistent with an
exceptionally ancient family-history. There are,
however, certain genera of ferns which afford
remarkable examples of restricted geographical
distribution associated with great antiquity. The

island of Juan Fernandez, 420 miles off the coast
of Chili, the home for four years of Alexander
Selkirk (to whose adventures we owe Defoe's
creation of Robinson Crusoe), is interesting also
from a botanical point of view. The vegetation of
this oceanic island, 20 square miles in area with
basaltic cliffs rising to a height of 3000 ft. above
the sea, includes more than 40 species of ferns,
eight of which occur nowhere else. One of these
endemic ferns is *Thyrsopteris elegans*, the only
representative of the genus; it is readily distinguished
by its large and graceful fertile fronds, examples of
which may occasionally be seen on a plant of this
species in the Royal Gardens at Kew : the sporangia
are produced in circular cups which replace the
ordinary leaflets on the lower branches of the frond
and hang from the short axis like miniature clusters
of grapes. It is noteworthy that among the fragmentary remains of the fern vegetation of the Jurassic
flora in England and in other parts of Europe
specimens occur with fertile segments practically
identical with those of the Juan Fernandez species.
Students of fossil plants are occasionally led away
by the temptation to identify imperfect specimens
with rare existing species to which they exhibit a
superficial resemblance, and this is well illustrated
by the frequent use of the generic name Thyrsopteris
for Jurassic and Lower Cretaceous ferns which are

too imperfect to be determined with any degree of certainty. We have, however, satisfactory grounds for the assertion that the Juan Fernandez fern affords a striking confirmation of the truth of Darwin's dictum that 'Rarity, as geology tells us, is the precursor to extinction.' In this remote oceanic island, for reasons which we cannot explain, there lingers an isolated type which belongs to another age.

The following passage, which forms a fitting introduction to an account of two other genera of ancient ferns, is taken from a description of an ascent of Mount Ophir in the Malay Peninsula by Dr A. R. Wallace in his well-known book on the Malay Archipelago :—'After passing a little tangled jungle and swampy thickets, we emerged into a fine lofty forest....We ascended steadily up a moderate slope for several miles, having a deep ravine on the left. We then had a level plateau or shoulder to cross, after which the ascent was steeper and the forest denser till we came out upon the Padang-Batu, or stone-field....We found it to be a steep slope of even rock, extending along the mountain side farther than we could see. Parts of it were quite bare, but where it was cracked and fissured there grew a most luxuriant vegetation, among which the pitcher plants were the most remarkable....A few coniferae of the genus Dacrydium here first appeared, and in the thickets, just above the rocky surface,

we walked through groves of those splendid ferns,
Dipteris Horsfieldii and *Matonia pectinata*, which

Fig. 11. *Matonia pectinata.* A group of plants in a wood on
Gunong Tundok, Mount Ophir. (Photograph by Mr A. G. Tansley.)

bear large spreading fronds on slender stems, 6 or
8 feet high '(49).

The two genera Matonia and Dipteris afford
exceptionally striking examples of survivals from
the past. Matonia is represented by two species,
Matonia pectinata (Fig. 11), which grows abundantly
on the upper slopes of Padang Batu in dense thickets
on the rock faces where, as Mr Tansley states, its
associates are a species of Gleichenia, Dipteris,
and a little *Pteridium aquilinum* (Bracken Fern).
Matonia pectinata occurs also on Bornean mountains
at an altitude of over 3000 ft. and descends to the
coast on some of the Malay islands. The other
species of the genus, *Matonia sarmentosa*, has so
far been found in one locality only, Niak, Sarawak,
where it was discovered by Mr Charles Hose.
Matonia pectinata has a creeping stem covered with
a thick felt of brown hairs bearing tall fan-shaped
fronds divided into numerous comb-like branches
thickly set with narrow linear leaflets on which
circular clusters of spore-capsules are sparsely scattered. In some respects Matonia is unlike other
ferns ; the fronds constitute a striking feature, and
the anatomy of the stem is still more distinctive.
In the form, development, and arrangement of the
sporangia (spore-capsules)—organs which from the
constancy of their characters have long been recognised as the most useful basis for classification—
Matonia exhibits distinctive features.

In order to emphasise the isolated position of the

genus it has recently been placed in a separate
family, the Matonineae, of which it is the sole living
representative. The restricted geographical range of
Matonia, considered in connexion with the clearly
marked peculiarities in structure and form, leads us
to expect other evidence in support of the natural
inference that the genus is a survivor of a once
more vigorous and widely spread family. If Matonia
were a recently evolved type which has not spread
far from its original home, we should expect it to
conform more closely than it does to other ferns in
the Malay region. Even assuming for the sake of
argument that variation may occur *per saltum*, and
new forms may be produced differing in more than
the finer shades of small variation from their parents,
the peculiar features of Matonia are too pronounced
and its individual characteristics too obvious to
warrant the assumption of recent production. It is,
however, from the testimony of the rocks that we
obtain confirmation of the opinion that these Malayan
species are plants on the verge of extinction. In
shales of Jurassic age exposed on the Yorkshire coast
at Gristhorpe Bay and in iron-stained rocks of the
same age between Whitby and Scarborough, well
preserved leaves have been found agreeing in the
shape of the frond, as also in the form of the leaflets
and of the groups of sporangia which they bare, with
those of *Matonia pectinata.*

The exposure by a stroke of the hammer, on the fractured surface of a rock picked up on the beach at Hayburn Wyke (a few miles south of Whitby), of a piece of fern frond which is unmistakably closely allied to the species described by Wallace on Mount Ophir, establishes a link between the Jurassic and the present era and presents a fascinating problem in geographical distribution. These fossil Matonias are known to students of ancient plants as species of the genus Matonidium, a name adopted by a German botanist for specimens apparently identical with those from the Yorkshire coast discovered in slightly younger rocks (Wealden) in North Germany. The same type has been found also in sediments of Wealden age on the Sussex coast. Other leaf-impressions agreeing closely with those of Matonidium have been obtained from the Yorkshire Jurassic rocks and these are assigned to another genus Laccopteris, an extinct member of the family Matonineae. It is not merely in the habit of the fronds and in the shape and venation of the leaflets that these fossil ferns resemble the existing species, but the more important features exhibited by the spore-capsules supply additional evidence. It has already been pointed out that the stems of Matonia are characterised by a type of structure unknown in an identical form in any other recent fern.

A few years ago Prof. Bommer discovered fragments of leaves and stems in Wealden beds a few miles from Brussels sufficiently well preserved to reveal the details of internal organisation. Some of these fossils were found to possess structural features identical with those of the Malayan species of Matonia. A full account of the fossil representatives of the Matonia family would be out of place in a general essay on Links with the Past, but brief reference may be made to some of the data which throw light on the geological history of the family. In strata classed by geologists as Rhaetic, a phase of earth-history between the Triassic and Jurassic eras (see p. 42), species of Laccopteris and allied forms have been described from several other countries ; from Jurassic and Wealden strata examples of both Laccopteris and Matonia have been found in Germany, Portugal, Belgium, Austria, and elsewhere. From rocks of Cretaceous age, higher in the series than the Wealden strata, well preserved impressions of a Matonidium have been discovered in Moravia. The Matonineae were widely distributed in Europe during the Rhaetic and Jurassic periods, but, so far as we know, the family did not survive in the northern hemisphere beyond the limits of the Cretaceous period. It is noteworthy that, in spite of the preservation of the remains of Jurassic and Cretaceous floras in many extra-European regions, notably in India, South Africa, Australia, China, and Tonkin, no specimens have been found which can

with confidence be assigned to the Matonineae.   A
single fossil has, however, been described from Queensland which may be a piece of a Laccopteris frond.

There is some evidence that ferns very similar to
Matonia existed in North America during the Mesozoic period.   It would be in the highest degree rash
to assume that the Matonineae played no part in
the Jurassic vegetation of India, South Africa, and
other southern lands, but there can be little doubt
that the family was especially characteristic of
European floras during a portion of the Mesozoic
era.   It would seem that subsequent to the Wealden
period the ancestors of Matonia dwindled in numbers
and their geographical range became much more
restricted.

The records of Tertiary rocks have hitherto added
nothing to our knowledge of the distribution of the
family subsequent to the Cretaceous period.   All
we can say is that the existing species of Matonia
are the last survivors of a family which in the
Jurassic period overspread a wide area in Europe
and probably extended to the other side of the
Atlantic. Exposed to unfavourable climatic conditions and possibly affected by the revolution in
the plant world consequent on the appearance of
the Flowering Plants, the Matonineae gradually
retreated beyond the equator until the two surviving
species found a refuge in the Malayan region.

Fig. 12. *Dipteris conjugata* Rein. and, in the middle of the upper
part of the photograph, a frond of *Matonia pectinata* R. Brown.
Mount Ophir. (Photograph by Mr A. G. Tansley.)

The fern spoken of by Dr Wallace as *Dipteris Horsfieldii* (perhaps better known as *Dipteris conjugata* (Fig. 12)), which grows with *Matonia pectinata* on Mount Ophir and in the Malay region generally, is one of seven species of a genus characterised by a somewhat wider geographical range than Matonia. *Dipteris conjugata* extends to the Philippines, Samoa, Fiji, New Caledonia, New Guinea and Central China; its fronds, like those of Matonia, are borne on long slender stalks attached to a creeping stem; they have a broad lamina divided by a deep median sinus into two symmetrical halves and each half is cut up into segments with a saw-like edge. Several stout ribs spread through the lamina from the apex of the long stalk like the open fingers of a hand; from these ribs smaller veins are given off at a wide angle, and these in turn give rise to a reticulum of finer veins forming a skeletal system like that in the leaves of an oak and many other flowering plants.

Numerous groups of spore-capsules are borne on the lower surface of the broad lobed frond. The leaves of other species of Dipteris have the same type of structure, but in some the segmentation of the lamina is carried further and the leaf consists of numerous long and narrow segments with one or two main ribs. Dipteris is represented in the flora of Assam, and it is interesting to find that a species

recently discovered in Borneo is more closely connected with the Assam type than with those of the
Malay region. Until a few years ago the genus
Dipteris was included in the large family Polypodiaceae of which nearly all our British ferns are
members, but the discovery of certain distinguishing
features in the structure of the sporangia showed
that these Eastern and Southern species form a
fairly well-defined group worthy of family rank.

In the Rhaetic plant-beds of Northern and
Central Europe, of North America, Tonkin, and
elsewhere, numerous fossil leaves have been discovered which in shape, venation, and in the
manner of occurrence of the sporangia bear a close
resemblance to species of Dipteris. Ferns of this
type were abundant in the Jurassic floras of the
northern hemisphere, and it is interesting to find
impressions of Dipteris-like leaves both in the
Jurassic rocks of the Yorkshire coast as well as in
slightly newer beds of the same geological period on
the north-east coast of Sutherland.

It is impossible to say with confidence how nearly
these Rhaetic and Jurassic ferns were related to the
existing species, as our knowledge of them is less
complete than in the case of the fossil representatives
of the Matonineae, but there can be no reasonable
doubt that in Dipteris as in Matonia we have a connecting link between the present and a remote past.

CHAPTER VI

THE REDWOOD AND MAMMOTH TREES
OF CALIFORNIA

'Your sense is sealed, or you should hear them tell
The tale of their dim life, with all
Its compost of experience.......' W. E. HENLEY.

Since their introduction into England about the
middle of the nineteenth century, the two Californian
species *Sequoia sempervirens* (the Redwood) and
Sequoia gigantea (the Mammoth tree) have become
familiar as cultivated trees. The name Sequoia,
said to be taken from Sequoiah, the inventor of the
Cherokee alphabet, was instituted in 1847, while the
name Wellingtonia, often used in horticulture though
discarded by botanists in favour of the older designation Sequoia, was proposed in 1853. Both species
are now confined to a comparatively small area in
California: their restricted geographical range, considered as an isolated fact, might be regarded as a
sign of recent origin. The records of the rocks,
however, afford ample proof that rarity in this as in

many other instances is the precursor of extinction.
The famous groves of Mariposa and Calaveras
represent the last resting-place of giant survivors
of a race which formerly held its own in Europe and
in other parts of the world.

The Redwood, *Sequoia sempervirens*, occupies a
narrow belt of country, rarely more than 20 or 30
miles from the coast, three hundred miles long from
Monterey in the south to the frontiers of Oregon;
it has a stronger hold on existence than *Sequoia
gigantea*. In Northern California it still forms pure
forests on the sides of ravines and on the banks of
streams. The tapering trunk, rising from a broad
base to over 300 ft., gives off short horizontal
branches thickly set with narrow spirally disposed
leaves $\frac{1}{4}$—$\frac{1}{2}$ inch in length arranged in two ranks
like the similar leaves of the Yew. The lower
edge of each leaf is decurrent, that is it runs a
short distance down the axis of the branch instead of
terminating at the point of attachment. It is by
paying attention to such details as this as well as to
more important features, that we are able to connect
fragmentary fossil twigs with those of existing
species. The female 'flowers' have the form of
oblong cones from $\frac{3}{4}$ to 1 inch long: each consists
of a central axis bearing crowded wedge-shaped,
woody appendages or cone-scales, which gradually
increase in breadth towards the exposed distal end

Fig. 13. *Sequoia gigantea.* King's Co., California.
(From Prof. D. H. Campbell)

characterised by its four sloping sides and by a
median transverse groove.   Several small seeds are
borne on the upper surface of the cone-scales.   The
smaller and short-lived male flowers need not be
described.

The other and better known species *Sequoia
gigantea* (Fig. 13) has an even more restricted range
and is confined to groves on the western slopes of
the Sierra Nevada between 3000 to 9000 ft. above
sea-level.

This tree is at once distinguished from the
Redwood by its ovate, sharply pointed and stiffer
leaves which retain their spiral disposition and
closely surround the axis of the twigs like obliquely-
set needles. The cones are of the same type as those
of *Sequoia sempervirens*, but are broader and may
attain a length of 3½ inches (9·5 cm.) (Fig. 14).

Reference has already been made to Sequoia
as a striking illustration of longevity.   It is also
selected as an equally impressive example of a type
verging on extinction, which played a prominent part
in the vegetation of both west and east during the
Cretaceous and Tertiary periods.

Scraps of branches with leaves hardly distinguishable from those of the existing Californian trees
are frequently met with in Tertiary and Mesozoic
sediments, and with them occasionally occur cones
too imperfectly preserved to afford satisfactory

Fig. 14. *Sequoia gigantea* Torr. ($\frac{2}{3}$ nat. size.)

7—2

evidence of more than superficial agreement with
those of the recent species.   The task of deciphering
the past history of plants, particularly of the Conifers,
is accompanied by many difficulties and insidious
temptations.   It is clear from a critical examination
of many of the recorded instances of fossil Sequoias
that the generic name has been frequently used by
writers without adequate grounds.   The fragmentary
specimens available to the botanical historian cannot
as a rule be subjected to microscopical investigation,
and even a partial acquaintance with the similarity
of the foliage of different types of living Conifers is
sufficient to convince the student of the need of self-
control in the identification of the fossils.   It is,
however, easy to point out obvious pitfalls, though
difficult to maintain a judicial attitude in the
excitement of endeavouring to interpret documents
which are too incomplete to be identified with
certainty.   If we put on one side all records of
supposed fossil Sequoias not based on satisfactory
data, there remains a wealth of material testifying to
the antiquity of the surviving species.

It is by no means improbable that Conifers
closely allied to the Redwoods and Mammoth trees
of California were represented in Jurassic floras ;
but hitherto no proof has been obtained of the
occurrence of a Sequoia among the rich material
afforded by the Jurassic plant-beds of Yorkshire

and by beds of the same age in other countries. A small cone has recently been described from strata near Boulogne belonging to the latest phase of the Jurassic period, which presents a strong resemblance in shape and size and in the form of the cone-scales to those of the recent species. This specimen, though not conclusive, is the most satisfactory indication of a Jurassic Sequoia so far discovered. From Lower Jurassic rocks in Madagascar similar cones have been recorded in association with foliage-shoots like those of *Sequoia gigantea*, but here too the evidence is not beyond suspicion. In plant-bearing strata of Wealden age, such as are exposed in the cliff near Hastings and in deposits of the same age in North Germany, Portugal, and elsewhere, twigs and cones have been found which may be those of trees nearly allied to the genus Sequoia.

It is, however, in the sedimentary rocks of Cretaceous age, rather higher in the series than those in the Hastings cliffs, and in the succeeding Tertiary rocks, that undoubted Sequoias are met with in abundance. At Bovey Tracey in Devonshire there is a basin-shaped depression in the granitic rocks of Dartmoor filled with clay, gravel and sand—the flood-deposits of a Tertiary lake containing waifs and strays of the vegetation on the surrounding hills. Among the commonest plants is one to which the late Oswald Heer gave the name *Sequoia Couttsiae*, and his reference of the specimens to the genus Sequoia has been confirmed by the recent researches of Mr and Mrs Clement Reid(50). This Tertiary (Oligocene) species is represented by slender twigs almost identical with those of *Sequoia gigantea* and by well-preserved cone-scales and seeds (Fig. 15).

A   B

Fig. 15.   *Sequoia Couttsiae* Heer.   Twigs (A) and cone-scales (B) from Bovey Tracey. (× 3.) (Photographs by Mr and Mrs Clement Reid.)

Moreover, it has been possible to examine micro-scopically the structure of the carbonised outer skin of the leaves and to demonstrate its agreement with that of the superficial tissue in the leaves of the Mammoth tree. With the Bovey Tracey Sequoia are associated fragments of Magnolia, Vitis, and *Taxodium distichum*, the swamp Cypress of North America, together with other types which have long ceased to exist in Western Europe. Other British examples of Sequoia have been described from Tertiary beds at Bournemouth, the Isle of Wight, Sheppey, and Antrim, but the material from these localities is inferior in preservation and cannot be identified with the same degree of certainty as in the case of the Devonshire specimens. The occurrence of twigs and cones of several species of Sequoia in both Cretaceous and Tertiary rocks in Austria, Germany, Italy, France, and elsewhere, shows that the ancestors of the Californian trees were common in the European region.

The exploration of Cretaceous and Tertiary rocks in Arctic Europe has revealed the former existence in Greenland, Spitzbergen, and other more or less ice-covered lands of plants which clearly denote a mild climate. Cones and branches of Sequoias have been found in abundance in Lower Tertiary beds on Disco Island off the west coast of Greenland, and similar evidence of the northern extension of the genus has been obtained from Spitzbergen. Dr Nathorst of Stockholm speaks of twigs of Sequoia in the Tertiary clays of Ellesmere Land almost as perfect as herbarium specimens. In Tertiary beds on the banks of the Mackenzie River, in Alaska, Saghalien Island and

Vancouver Island, and in Upper Cretaceous rocks in the Queen Charlotte Islands, remains of Sequoia have been discovered. One of the most remarkable instances of the preservation of trees of a bygone age is supplied by the volcanic deposits of Lower Tertiary age exposed on the slopes of Amethyst mountain in the Yellowstone Park district. At different levels in the volcanic and sedimentary material, which is piled up to a height of over 2000 ft. above the valley, as many as fifteen forests are represented by erect and prostrate limbs of petrified trees (Fig. 6). The microscopical examination of some of these trees has shown that they bear a striking resemblance to *Sequoia sempervirens*. In a photograph of these petrified forests by the U.S. Geological Survey [36, 2] one sees living Conifers side by side with the lichen-covered and weathered trunks of the fossil species (*Sequoia magnifica*), living and extinct being at a distance hardly distinguishable.   (Frontispiece.)

In concluding this brief survey of the fossil records of Sequoia, reference may be made to the discovery of petrified wood in Cretaceous rocks in South Nevada, possessing the anatomical features of *Sequoia gigantea*, which shows that close to the present home of the big trees their ancestors flourished during a period of the earth's history too remote to be measured by human reckoning.

The distribution of the Tertiary and Cretaceous

Sequoias would appear to have been mainly in the northern hemisphere, extending well within the Arctic circle.   It is, however, by no means improbable that the ancestors of Sequoia flourished far south of the equator.   Reference has been made to Jurassic fossils from Madagascar which have been compared with the existing species, and from Lower Tertiary beds in New Zealand the late Baron Ettingshausen described some cones and twigs as *Sequoia novae zeelandicae* which bear a close resemblance to the existing type. The available evidence would seem to point to a northern origin of the genus, though allowance must be made for erroneous conclusions based on negative evidence.   Further research may well extend the past distribution of Sequoia in southern lands, but the data to hand point to the conclusion that the Californian trees represent the survivors of a type which flourished in the Cretaceous and Tertiary periods over a wide area in North America and in what we now call the Continent of Europe.

CHAPTER VII

THE ARAUCARIA FAMILY

And so the grandeur of the Forest-tree
Comes not by casting in a formal mould,
But from its *own* divine vitality.' WORDSWORTH

As an additional illustration of existing cone-
bearing trees which form links with the past we may
briefly consider the genera Araucaria and Agathis,
the two members of the family Araucarieae. It is
generally agreed that the branches of the genealogical
tree of this family extended farther back into the
past than in the case of the majority of Conifers. By
some authors the surviving representatives of the
Araucarian stock are considered to have a strong
claim to be regarded as the most primitive as well as
the oldest of cone-bearing trees, though this opinion,
like many others, is not held by botanists as a whole.
This is not the place to discuss matters of controversy,
and I shall confine myself to a general consideration
of Araucaria and Agathis from the point of view of
their present distribution and the part they played in

the vegetation of the Mesozoic and Tertiary epochs. In 1741 a plant from Amboyna, one of the Moluccas, was described under the name *Dammara alba*. For this tree, known as the Amboyna Pine, the English botanist Salisbury instituted the generic name Agathis, from a Greek word (ἀγαθίς) meaning a ball of string and probably suggested by the form of the cones, which is the designation usually adopted in botanical literature instead of the pre-Linnean term Dammara. The best known species of the genus is the Kauri Pine, probably the finest forest tree in New Zealand where it still flourishes from the North Cape to latitude 38° S., though the occurrence of sub-fossil trunks and pieces of buried resin shows that the Kauri forests are gradually dwindling. The stems of this species, *Agathis australis*, rise like massive grey columns to a height of 160 ft., terminating in a succession of spreading branches given off in tiers from the main trunk. The thick narrow lanceolate leaves, with several parallel veins, reach a length of 2 to 3 inches. The female shoots have the form of small and almost spherical cones consisting of a central axis bearing overlapping spiral series of broadly triangular scales (Fig. 16). Each scale carries a single seed with a large wing attached to one side which facilitates disposal by wind. Other species of Agathis occur in the Malay Archipelago, the Philippines, in Queensland, in the New Hebrides, New Caledonia, the Fiji Islands, and

Fig. 16.   A. *Agathis robusta* Muell. (much reduced).   B.   *Agathis Moorei* Lind. (½ nat. size).

elsewhere   With the exception of the Australian Kauri (*Agathis robusta*), with leaves larger and broader than those of the New Zealand Kauri, the genus is essentially an island type.   With the exception of some species of the southern hemisphere genus Podocarpus, there are no Conifers with foliage like that of Agathis.   It is, however, the broad and thin single-seeded scales and the spherical cones, in some species six inches in length, which furnish the most trustworthy means of identifying the genus.

The allied genus Araucaria, with the exception of two South American species, the familiar Monkey Puzzle, *Araucaria imbricata*, and a Brazilian tree, *Araucaria brasiliana*, is confined within the geographical area occupied by Agathis.   The name Araucaria was first used by de Jussieu in 1789 for a plant previously referred to the genus Pinus and described as one of the most beautiful trees of Chili.   This species, *A. imbricata*, introduced into England in 1796, grows on the southern slopes of the Andes and, as in the case of the Kauri forests of New Zealand, buried stems point to a wider extension of the forests in earlier days.   The sharp and thick leaves of the Monkey Puzzle distinguish it from all other Conifers; its large almost spherical seed-bearing cones, more than half a foot in length, which may occasionally be seen on well-grown British trees, are unlike those of other genera.   Each of the deep and narrow scales bears a single seed embedded in the substance of the scale and terminates distally in a narrow upturned process. Some species of Araucaria, differing considerably in the form of the leaves and in the shape and structure of the seed-scales from the Chilian species, are conveniently placed in a distinct sub-

Fig. 17.   *Araucaria excelsa*.   The upper part of a small tree in the Cambridge Botanic Garden.   (Much reduced.)

division of the genus Araucaria.   Of this type the Norfolk Island Pine, *Araucaria excelsa*, is the best-known example (Fig. 17).   It was introduced to Kew by Sir Joseph Banks in 1793, soon after its discovery by Captain Cook, who describes the stems of the

Norfolk Island trees as resembling basaltic columns, and relates how on approaching the island everyone was satisfied that the columnar objects were trees, except our Philosophers, who still maintained they were basaltes.' The leaves are short, about half an inch long, laterally compressed and slightly spreading and sickle-shaped—sometimes shorter and broader and overlapping—arranged in crowded spirals. The scales of the broadly oval cones are single-seeded, but differ from those of *Araucaria imbricata* in having the seed exposed on the surface and in the greater breadth and thinner borders of the scales. In both Araucaria and Agathis the nature of the seed-scales constitutes a distinguishing feature. The leaves of *Araucaria imbricata* differ in form from those of other Conifers. The foliage shoots of *Araucaria excelsa* and other species, *e.g.* the very closely allied *A. Cookii* of the New Hebrides and New Caledonia, though not unlike the branches of a Japanese Conifer (*Cryptomeria japonica*), often cultivated in England, afford fairly trustworthy characters for identification purposes.

The minute structure of the wood of both Araucaria and Agathis constitutes an important distinguishing feature and enables us to recognise on microscopical examination even a fragment of wood of either of these genera. The small elongated cells or water-conducting elements of the wood of the Araucarieae are characterised by one or two, and occasionally as many as three or four, contiguous rows of pits on their radial walls, and these appear in surface view as flattened circles or polygonal areas.

These details have been mentioned in order to show that Araucaria and Agathis are sufficiently distinct in many respects from other Conifers to render their identification in a fossil state comparatively easy, at least much easier than the recognition of the majority of the members of the Coniferae. It would be going too far to state definitely that Araucarieae, as defined by reference to existing species, existed during the Palaeozoic period ; on the other hand it would seem in a high degree probable that the vegetation of the Coal age and of the succeeding Permian period included trees in which certain Araucarian characters were clearly foreshadowed. The name Araucarioxylon was formerly applied to petrified wood, obtained from Palaeozoic as well as from later formations, which agrees anatomically with that of Araucaria and Agathis. It has been shown in recent years that much of the Palaeozoic wood of this type of structure belongs to the extinct genus Cordaites, a tree which played a prominent part in the earlier floras. Cordaites affords a good example of a generalised type : in its wood-structure it resembles very closely the existing Araucarieae ; its long strap-like leaves are not unlike those of some species of Agathis ; its male flowers have often been compared with those of the Maiden Hair tree, *Ginkgo biloba,* and certain anatomical features form connecting links between this Palaeozoic genus and the Cycads.

It is noteworthy that in another Palaeozoic genus, Walchia, the leaf-bearing branches are identical in appearance with those of the Norfolk Island Pine (Fig. 17) and some other species of Araucaria. Unfortunately our knowledge of the reproductive organs of Walchia is insufficient to warrant any definite statement as to the degree of consanguinity between this Permian and Upper Carboniferous plant and the Araucarieae ; it is probable that in Walchia we have a type not far removed from the line of evolution which led to Araucaria. Petrified wood, identified as that of Walchia, and exhibiting the Araucarian type of structure, has been recorded from Permian rocks of the Vosges. Other instances might be quoted in support of the view that the Palaeozoic floras included a few plants with which the surviving Araucarieae may fairly claim relationship. Professor Zeiller of Paris has recently described some fossil shoots from Palaeozoic rocks in India under the name *Araucarites Oldhami* on the ground of the similarity of the leaves to those of *Araucaria imbricata.* Similarly, from Triassic rocks several fossils have been described as closely allied to Araucaria, in some cases because of anatomical resemblances and in others on the less satisfactory evidence furnished by a similarity in the

s.   8

foliage shoots.   Professor Jeffrey of Harvard has recently given an account of a new type of stem
(Woodworthia) from the petrified Triassic forest of
Arizona possessing some Araucarian characters,
though differing from existing species of Araucaria
in certain structural features, a combination of
characters regarded by this Author as an indication
of relationship with the family of Conifers, which
includes the Pines, Firs, Larches and other well-known
northern genera.

It is, however, from the records of Jurassic rocks
that we obtain the most satisfactory information as to
the great antiquity and the very wide geographical
range of the ancestors of the recent genus. The plant-
beds of the Yorkshire coast afford clear evidence of
the occurrence of Araucarian trees in the woodlands
of the Jurassic period. Petrified wood has been found
at Whitby, associated with jet, showing the minute
structural characteristics of the surviving species of
Araucarieae, and it is not improbable that some at
least of the Whitby jet has been formed from the wood
of Araucarian plants. The carbonised remains of
leafy shoots preserved in the Jurassic shales near
Scarborough and on other parts of the Yorkshire
coast include twigs hardly distinguishable from those
of *Araucaria excelsa*, though the resemblance of
external form alone, especially in the case of foliage
shoots, does not amount to proof of generic identity.

We have, however, the much more trustworthy evidence of cones and seed-bearing scales in which the characteristic features of living species are clearly shown. Seed-bearing scales almost identical with those of *Araucaria excelsa* and other recent species have long been known from the Jurassic rocks of Yorkshire.

From other parts of England where samples of Jurassic floras are preserved, as at Stonesfield in Oxfordshire, in Northamptonshire and elsewhere, equally striking examples of undoubted Araucarias have been found.

Fig. 18 represents part of a large cone described in 1866 by Mr Carruthers from Jurassic rocks at Bruton in Somersetshire : this specimen, now in the British Museum, consists of one side of a spherical cone about 5 inches long and 5 inches broad; in size, as in the form of the seed-scales, it shows a striking likeness to the cones of the Australian species *Araucaria Bidwillii*, the Bunya Bunya of Queensland. Other equally convincing examples of Jurassic Araucarian cones and seeds may be seen in the museums of York and Northampton. On the northeast coast of Sutherland there is a narrow strip of Jurassic beds forming a low platform between the granitic and Old Red Sandstone hills and the sea. From these rocks Hugh Miller described several fossil plants in his *Testimony of the Rocks*, and an examination of a large collection obtained from this

8—2

Fig. 18. *Araucarites sphaerocarpus* Carr. From Jurassic rocks at
Bruton, Somersetshire. (British Museum. $\frac{2}{3}$ nat. size.)

district by the late Dr Marcus Gunn shows that
Miller was justified in speaking of Araucaria as a
member of this northern flora.

There is abundant evidence pointing to the existence in Britain during the Jurassic period, and in
the early days of the Cretaceous epoch, of Araucarian
trees which differed but slightly from the modern
species confined to the southern hemisphere.   In
several localities in France, Germany, and other parts
of the continent, Araucarian fossils have been recognised in Jurassic rocks.   It is almost certain that
some foliage shoots and imperfectly preserved cones
described by Dr Nathorst from Upper Jurassic rocks
in Spitzbergen were borne by a species of Araucaria.
Cone-scales very similar to those from Yorkshire have
been discovered in Wealden beds in Cape Colony, and
Araucarian wood of Jurassic and Cretaceous age has
been found in Madagascar.   From Jurassic strata in
India and Victoria (Australia), as well as from Upper
Jurassic and Lower Cretaceous rocks in Virginia and
elsewhere in the eastern United States, well preserved
Araucarian fossils are recorded.   In a collection of
Jurassic plants, obtained a few years ago by the
members of a Swedish Antarctic Expedition in
Graham's Land, Dr Nathorst has recognised some
cone-scales of Araucaria, which demonstrate a former
extension of the family beyond the southern limits of
South America.

It is interesting to find that when we ascend higher in the geological series and pass beyond the Wealden strata to the Middle and Upper sub-divisions of the Cretaceous period, evidence of the wide geographical distribution of the Araucarieae is still abundant. Araucarian wood has been obtained in rocks classed as Upper Cretaceous in Egypt, in East Africa, in Dakota, and elsewhere. In the sedimentary rocks of the Tertiary period undoubted examples of Araucaria are less common, though there can be no doubt that the genus was much more widely spread then than it is at the present day. The well-known Tertiary plant-beds of Bournemouth have afforded specimens of foliage shoots which have been described as a species of Araucaria, though in the absence of well-preserved cones or petrified wood we must admit that the data are inconclusive. It is, however, legitimate to regard the striking similarity of the Bournemouth twigs to those of *Araucaria excelsa* and *A. Cookii* as constituting a fairly strong case in favour of the persistence of Araucaria in Western Europe up to the earlier stage of the Tertiary period. Araucarian wood of Tertiary age is recorded from India, while branches with broad leaves like those of *Araucaria imbricata* have been found in Seymour Island and the Magellan Straits, and specimens of Tertiary wood are described from Patagonia. At the other end of the world, Tertiary rocks on the

west coast of Greenland have yielded fragments which
may be referred with some hesitation to the genus
Araucaria.

A few words must be added in regard to the recent
discovery by Professor Jeffrey and Dr Hollick of some
very interesting Cretaceous specimens in New Jersey
of well-preserved cone-scales and foliage shoots of
extinct plants closely related to the existing species
of Agathis (51).   The American fossils are particularly
valuable because their preservation admits of microscopical examination of the tissues.   In Cretaceous
rocks of Staten Island and in other localities on the
eastern border of the northern United States, kite-
shaped seed-bearing scales almost identical in form
with those of recent species of Agathis are fairly
common fossils.   Similar specimens have long been
known from Tertiary rocks in western Greenland.
In the case of some of the American examples each
scale bore three seeds instead of a single seed in
living species: on account of this difference Prof.
Jeffrey and Dr Hollick have adopted a distinct
generic name, *Protodammara.*

The foregoing sketch is necessarily far from complete, but it may serve as an illustration of the light
which is thrown on the past history of recent plants
by the investigation of the relics of ancient floras.
The family Araucarieae now represented by a small
number of species which, with the exception of the

Andian and Brazilian Araucarias, are restricted to a
small region in the southern hemisphere, was one of
the most widely spread sections of the seed-bearing
plants during the Mesozoic era.   Ancestors of Araucaria must have been common trees in the European
vegetation in Jurassic and Lower Cretaceous periods,
and even as late as the Tertiary period there is evidence
that representatives of the family still lingered in
the north.   One conclusion which seems almost un-
avoidable is that the species of Araucaria and Agathis
that survive, in some cases only in one or two small
islands in the South Pacific, have in the course of
successive ages wandered from the other end of the
world.   Their migrations can be partially traced by
the fragments embedded in Jurassic and later sediments, but we can only speculate as to the causes
which have contributed to the changes in the fortunes
of the family, how much influence may have been
exerted by changes in physical conditions in the
environment, and to what extent the production of
more successful types may have been the dominant
cause of the decline, it is impossible to say.   One
thing at least is certain, that few existing plants are
better entitled to veneration as survivals from the
past than are the living species of Araucaria.

CHAPTER VIII

THE MAIDEN HAIR TREE

'...the trees
That whisper round a temple become soon
Dear as the temple's self.' KEATS.

The Maiden Hair tree of China and Japan, which was introduced into Europe early in the eighteenth century, has now become fairly well known. Though hardy in England, it requires warmer summers for full development and regular flowering. To botanists this Eastern tree is of peculiar interest, partly because of the isolated position it occupies in the plant-kingdom and partly by reason of its great antiquity. There is probably no other existing tree which has so strong a claim to be styled a 'living fossil,' to use a term applied by Darwin to survivals from the past. In 1712 the traveller Kaempfer proposed for this plant the generic name Ginkgo, and Linnaeus adopted this designation, adding the specific name *biloba* to denote the bisection of the wedge-shaped lamina of the leaf into two

divergent segments.   In 1777 the English botanist
Sir J. E. Smith expressed his disapproval of what

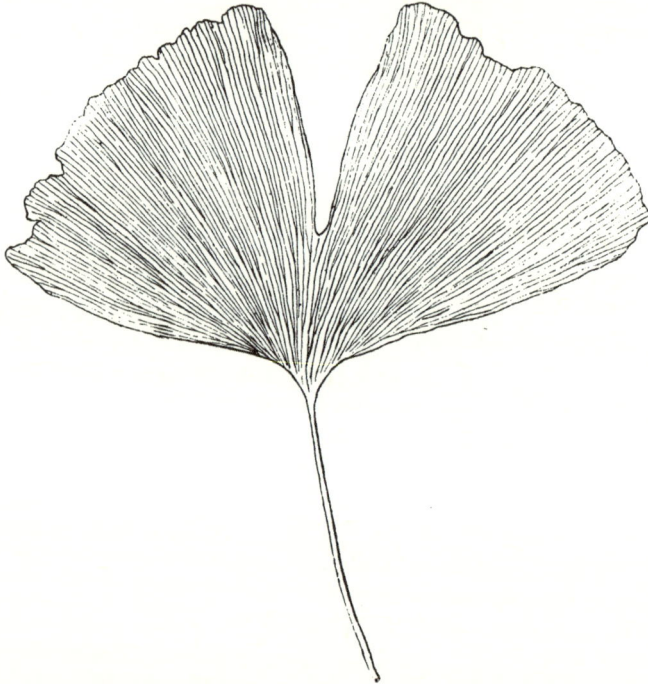

Fig. 19.   *Ginkgo biloba* Linn.   (Slightly reduced.)

he called the uncouth name Ginkgo by substituting
for *Ginkgo biloba* the title *Salisburia adiantifolia,*

but as it is customary to retain names adopted or
proposed by Linnaeus, the founder of the binominal
system of nomenclature, the correct botanical designation of the maiden hair tree is *Ginkgo biloba*.
Mere personal preference such as that of Sir J. E.
Smith for Salisburia is not an adequate reason for
rejecting an older name.

In its pyramidal habit Ginkgo agrees generally
with the larch and other Conifers. Like the larch
and cedar it possesses two kinds of foliage shoots,
the more rapidly growing long shoots with scattered
leaves and the much shorter dwarf-shoots which
elongate slightly each year and bear several leaves
crowded round their apex.   The leaves (Fig. 19),
which are shed each year, are similar in the cuneate
form of the lamina and in the fan-like distribution
of the forked veins, to the large leaflets of some
species of maiden hair ferns : the thin lamina carried
by a slender leaf-stalk is usually about 3 inches
across, though in exceptional cases it may reach a
breadth of 8 inches. The lamina is usually divided
by a deep V-shaped sinus into two equal halves ;
it may be entire with an irregularly crenulate margin,
or, on seedlings and vigorous long shoots, the lamina
may be cut into several wedge-shaped segments.

The male and female flowers are borne on separate
trees ; the male consists of a central axis giving off
slender branches, each of which ends in a small

terminal knot and two elliptical capsules in which
the pollen is produced. The female flowers have
a stouter axis which normally produces two seeds
at the apex. The seed is encased in a green fleshy
substance and, as in the fruit of a cherry or plum,
the kernel is protected by a hard woody shell. In
the form of the leaves and in the structure of the
flowers Ginkgo presents features which clearly distinguish it from the Conifers, the class in which,
until recently, it was included. In 1896 the Japanese
botanist Hirase made the important discovery that
the male reproductive cells of Ginkgo are large
motile bodies provided with a spirally coiled band
of minute cilia—delicate hairs which by their
rapid lashing-movement propel the cell through
water. In all Flowering Plants and in Conifers the
male reproductive cells have no independent means
of locomotion ; they are carried to the female cell
by the formation of a slender tube—the pollen-tube—
produced by the pollen-grain. In the Ferns, Lyco-
pods and Horsetails—in fact in all members of
the Pteridophyta—as also in the Mosses and Liverworts as well as in many of the still lower plants, the
male cells swim to the egg by the lashing of cilia
like those on the male cells of Ginkgo. This difference
in regard to the nature of the male cells was considered to be a fundamental distinction between the
higher seed-bearing plants and all other groups of

the vegetable kingdom. It was, therefore, with no
ordinary interest that Hirase's discovery was received,
as it broke down a distinction between the two great
divisions of the plant-world which had been generally
accepted as fundamental ; though it is only fair to
say that the German botanist Hofmeister, a man of
exceptional originality and power of grasping the
essential, foresaw the possibility that this arbitrary
barrier would eventually be removed.   The Ferns
and other plants in which the male cells are motile,
represent earlier stages in the progress of plant
development, when the presence of water was essential for the act of fertilisation, a relic of earlier days
when the whole plant-body was fitted for a life in
water.   As higher types were produced, the plant-
machinery became less dependent on an aqueous
habitat, and the loss of organs of locomotion in the
male cells is an instance of the kind of change
accompanying the gradual adaptation to life on
land.   The idea of the gradual emancipation of plants
from a watery environment is expressed in a somewhat extreme form by the author of a book entitled
The Lessons of Evolution (52), who states that the
ocean is the mother of plant-life and that plants
formed the army which conquered the land.   In
Ginkgo we have a type which, though similar in
most respects to the Conifers, possesses in its motile
reproductive cells a persistent inheritance from the

past. The recognition of this special feature afforded
a sound reason, especially when other peculiarities are
considered, for removing Ginkgo from the Conifers
and instituting a new class-name, Ginkgoales.

Ginkgo is a generalised type, linked by different
characters both with living members of the two classes
of naked-seeded plants and with certain existing
Palaeozoic genera. It is a survivor of a race which
has narrowly escaped extinction; the last of a long
line that has outlived its family and offers by its persistence an impressive instance of the past in the
present. Though Mrs Bishop in her *Untrodden
Paths in Japan* speaks of forests of Maiden Hair
trees apparently in a wild state, it is generally
believed that they were cultivated specimens.
Mr Henry who has an exceptionally wide knowledge
of Chinese vegetation tells us that 'all scientific
travellers in Japan and the leading Japanese botanists
and foresters deny its being indigenous in any part of
Japan; and botanical collectors have not observed it
truly wild in China.' Moreover, Mr E. H. Wilson,
after traversing the whole of the district where
Ginkgo was supposed to occur in a wild state, says
that he found only cultivated trees. There is no reason
to doubt that China is the last stronghold of this
ancient type which in an earlier period of the earth's
history overspread the world.

A brief summary of the past history of Ginkgo

and of the Ginkgoales supplies overwhelming testimony to the tenacity of life with which the Maiden Hair tree has persisted through the ages.

It was pointed out in the account of the past history of Araucaria that the records obtained from Palaeozoic rocks, while affording evidence of the existence of Carboniferous and Permian genera undoubtedly allied to the living species, do not enable us to speak with certainty as to the precise degree of affinity. Similarly, Palaeozoic leaves have been described as representatives of the class of which Ginkgo is the sole survivor, but the evidence on which this relationship is assumed is by no means conclusive.

The generic name Psygmophyllum has been applied to some impressions of Ginkgo-like leaves discovered in the Upper Devonian rocks of Spitzbergen, a small remnant of land in the Arctic circle, which has furnished valuable information as to the composition of one of the oldest floras of which satisfactory remains have been found. Other examples of these lobed, wedge-shaped leaves are recorded from Carboniferous rocks in Germany, France, and elsewhere ; from Permian strata in the east of Russia and from Palaeozoic beds in Cape Colony and Kashmir. A relationship between Psygmophyllum and Ginkgo is, however, by no means established and rests solely on a resemblance in the form of the leaves. The close correspondence in form and venation between some leaves from Permian rocks in the Ural mountains and from Lower Permian beds in France, and those of the recent species, is considered by some authors sufficiently striking to justify the reference of these fossils to the genus Ginkgo. Similar leaves of Permian age, which may also be related to the existing species, have been described under the name Ginkgophyllum. Other specimens of Palaeozoic age from North America and elsewhere have been assigned to the Ginkgoales; but in none of these cases, despite the resemblance in leaf-form, is there sufficiently convincing evidence of close relationship to warrant a definite assertion that the plants in question were members of the group of which Ginkgo alone remains.

It is, however, an undoubted fact that the Maiden Hair tree is connected by a long line of ancestors with the earliest phase of the Mesozoic era. From many parts of the world large collections of fossil plants have been obtained from strata referred to the Rhaetic period, or to the upper division of the Triassic system. A comparison of floras from these geological horizons in different parts of the world points to a vegetation extending from Australia, Cape Colony, and South America, to Tonkin, the south of Sweden and North America, which was character- ised by a greater uniformity than is shown by widely separated floras at the present day. One of the com- monest genera in Rhaetic floras is that known as

Baiera ; this name is applied to wedge-shaped leaves with a slender stalk similar in shape and venation to those of Ginkgo, but differing in the greater number and smaller breadth of the segments. Between the deeply dissected leaf of a typical Baiera with its narrow linear lobes and the entire or broadly lobed leaf of a Ginkgo there are many connecting links, and to some specimens either name might be applied with equal fitness. Examples of Baiera leaves, in some cases associated with fragments of reproductive organs, are recorded from Rhaetic rocks of France, the south of Sweden, Tonkin, Chili, the Argentine, North America, South Africa, and from other regions. There is abundant evidence pointing to the almost world-wide distribution of the Ginkgoales, as represented more especially by Baiera, in the older Mesozoic floras. In the later Jurassic rocks of Yorkshire true Ginkgo leaves as well as those of the Baiera type are fairly common ; with the leaves have been found pieces of male and female flowers. Ginkgo and Baiera have been described from Jurassic rocks of Germany, France, Russia, Bornholm, and elsewhere in Europe ; they occur abundantly in Middle Jurassic rocks in northern Siberia, and are represented in the Jurassic floras of Franz Josef Land, the East Coast of Greenland, and Spitzbergen (Fig. 20). The abundance of Ginkgo and Baiera leaves associated with male flowers and seeds discovered in Jurassic rocks,

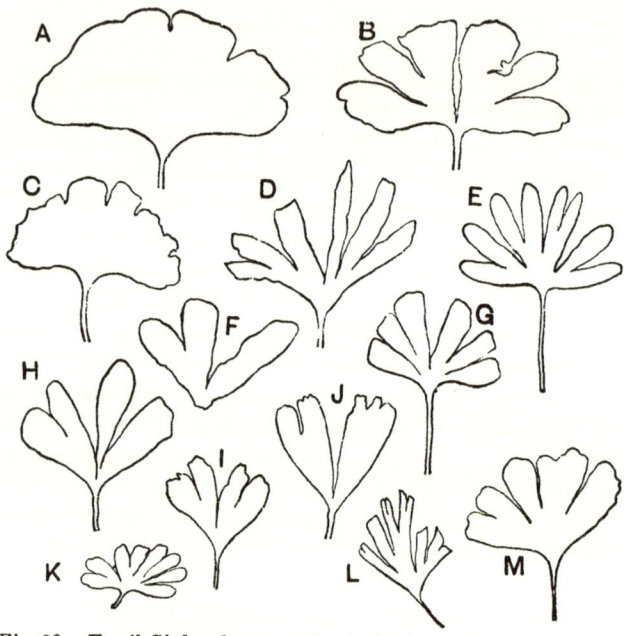

Fig. 20. Fossil Ginkgo leaves. (½ nat. size.)
   A. Tertiary, Island of Mull.
   B. Wealden, North Germany (after Schenk).
   C. Jurassic, Japan (after Yokoyama).
   D. Jurassic, Australia (after Stirling).
   E. Jurassic, Siberia (after Heer).
   F. Jurassic, Turkestan.
   G. Lower Cretaceous, Greenland (after Heer).
   H. Jurassic, California (after Fontaine).
   I. Jurassic, Yorkshire.
   J. Jurassic, N.E. Scotland (after Stopes).
   K. Wealden, Franz Josef Land (after Nathorst).
   L. Rhaetic, South Africa.
   M. Jurassic, Spitzbergen (after Heer).

approximately of the same geological age as those on the Yorkshire coast, in East Siberia and in the Amur district, has led to the suggestion that this region may have been a centre where the Ginkgoales reached their maximum development in the Mesozoic period.

It should be added that other genera of Jurassic and Rhaetic fossils in addition to Ginkgo and Baiera have been referred to the Ginkgoales, though evidence of such affinity is not convincing. There is, however, good reason to believe that this widespread group was represented by several genera in the older Mesozoic floras.

The occurrence of the Ginkgoales in Jurassic rocks in King Charles Land and in the New Siberian Islands (lat. 78° and 75° N.), in Central China, Japan, Turkestan, California, Oregon, South Africa, Australia, and Graham's Land demonstrates the cosmopolitan nature of the group. During the later part of the Jurassic period and in the Wealden floras both Baiera and Ginkgo were abundant; leaves are recorded from Jurassic strata in the north-east of Scotland, from Lower Cretaceous or Wealden rocks in North Germany, Portugal, Vancouver Island, Wyoming, and Greenland.

During the Tertiary period, or probably in the earlier days of that era, Ginkgo flourished in North America, in Alaska and in the Mackenzie River district, Greenland, Saghalien Island, and in several European regions. In Chapter III reference was made to the

9—2

volcanic activity which characterised the north-west
European area in the early Tertiary period and resulted
in the formation of the thick sheets of basalt on the
north-east coast of Ireland and in the Inner Hebrides.
There were occasional pauses in the volcanic activity,
during which vegetation established itself on the
weathered surface of the lava, and left traces of its
existence in the leaves and twigs preserved in the
sedimentary material enclosed between successive
lava-floras. At Ardtun Head in the Isle of Mull
beautifully preserved leaves of Ginkgo, 2—4 inches
in breadth, with the median sinus and the venation
characteristic of the leaves of the existing plant, have
been discovered in a bed of clay which marks the site
of a lake in a depression on the lava-plateau.   The
resemblance of these Tertiary leaves from Mull to
those of the surviving Maiden Hair tree is so close as
to suggest specific identity. Mr Starkie Gardner and
Baron Ettingshausen have described some seeds from
the London clay (Lower Tertiary) in the Isle of Sheppey
as those of Ginkgo, but this identification rests on
data too insufficient to be accepted without hesitation.

The recent cultivation of *Ginkgo biloba* in Britain
may therefore be spoken of as the re-introduction of
a plant which in the earlier part or in the middle of
the Tertiary period flourished in the west of Scotland,
and was abundant in England in the earlier Jurassic
period. It is impossible to say with any confidence
where the Ginkgoales first made their appearance,

whether in the far north or in the south, nor are we able to explain the gradual decline of so venerable and vigorous a race.

As we search among the fragmentary herbaria scattered through the sedimentary rocks in that comparatively small portion of the earth's crust which is accessible to investigation, we discover evidence of a shifting of the balance of power among different classes of plants in the course of our survey of successive floras. Plants now insignificant and few in number are found to be descendants of a long line of ancestors stretching back to a remote antiquity when they formed the dominant class. Others which flourished in a former period no longer survive, either themselves or in direct descendants. 'The extinction of species has been involved in the most gratuitous mystery.' We can only speculate vaguely as to the cause of success or failure. Certain types were better armed for the struggle for life, and produced descendants able to hold their own and to perpetuate the race through the ages in an unbroken line. Others had a shorter life and fell out of the ranks of the advancing and ever changing army. To quote Darwin's words: 'We need not marvel at extinction; if we must marvel, let it be at our own presumption in imagining for a moment that we understand the many complex contingencies on which the existence of each species depends.'

BIBLIOGRAPHY

Many of the books and papers dealing with subjects touched upon in this volume are not included in the following list. For reference to a more complete bibliography the reader should consult more technical treatises.

1. HOLMES, T. RICE. Ancient Britain and the invasions of Julius Caesar. Oxford, 1907.
2. MITCHELL, A. The Past in the Present; what is Civilisation? Edinburgh, 1880.
3. WEISMANN, A. Essays upon Heredity and Kindred Biological Problems. (I. The Duration of Life.) Vol. I. Edited by E. B. Poulton, S. Schönland, and A. E. Shipley. (Second Edition.) Oxford, 1891.
4. THE HISTORIE OF THE WORLD, commonly called the Naturall Historie of the C. PLINIUS SECUNDUS. Translated into English by Philemon Holland. London, 1634.
5. HOOKER, J. D. On Three Oaks of Palestine. *Trans. Linnean Society*, Vol. XXIII. p. 381. 1862.
6. HOLTERMANN, CARL. Der Einfluss des Klimas auf den Bau der Pflanzengewebe. Leipzig, 1907.
7. ATKINSON, A. Notes on an Ancient Boat found at Brigg. *Archaeologia*, Vol. L. p. 361. 1887.
8. HUXLEY, T. H. Man's Place in Nature and other Anthropological Essays. *Collected Essays*, Vol. VII. (On the methods and results of Ethnology.) London, 1901.
9. LEWIS, F. J. The sequence of Plant Remains in the British Peat Mosses. *Science Progress*, No. 6. October, 1907.

10. STRAHAN, A. On submerged Land-surfaces at Barry, Glamorganshire. With notes on the Fauna and Flora by Clement Reid; etc. *Quart. Journ. Geological Society*, Vol. LII. p. 474. 1896.

11. THE LIFE AND LETTERS OF CHARLES DARWIN. Edited by Francis Darwin. 3 Vols. London, 1887.

12. MORE LETTERS OF CHARLES DARWIN. Edited by Francis Darwin and A. C. Seward. 2 Vols. London, 1903.

13. WARD, LESTER F. The Course of Biologic Evolution. Anniversary Address of the President of the Biological Society. Washington, 1890.

14. MARSHALL, W. *Anacharis alsinastrum*, a new water weed. (Reprinted from the Cambridge *Independent Press*.) London, 1852.

15. BAILEY, C. Notes on the structure, the occurrence in Lancashire, and the source of origin, of *Naias gramineus* Delile, var. *Delilei magnus*. *Journal of Botany*, Vol. XXII. p. 305. 1884.
WEISS, F. H. and H. MURRAY. On the occurrence and distribution of some alien aquatic plants in the Reddish Canal. *Mem. Proc. Manchester Lit. and Phil. Society*, Vol. LIII. Pt. II. 1909.

16. BENNETT, A. The Halifax Potamogeton. *Naturalist*, No. 621. October, 1908.

17. HOOKER, J. D. Outlines of the Distribution of Arctic Plants. *Trans. Linn. Soc.* Vol. XXIII. p. 251. 1862.

18. ENGLER, A. Plants of the Northern Temperate Zone in their transition to the High Mountains of Tropical Africa. *Annals of Botany*, Vol. XVIII. p. 523. 1904.

19. DARWIN, C. The Origin of Species. London, 1900.

20. RIDLEY, H. N. On the dispersal of seeds by wind. *Annals of Botany*, Vol. XIX. p. 351. 1905.

21. HOOKER, J. D. On the Cedars of Lebanon, Taunus, Algeria, and India. *The Natural History Review*, 1862, p. 11.

22. FORBES, E. On the connection between the distribution of the existing Fauna and Flora of the British Isles, and the geological changes which have affected their area.... *Memoirs, Geological Survey*, Vol. I. p. 336. 1846

23. PRAEGER, R. L. The Wild Flowers of the West of Ireland and their history. *Journ. R. Hort. Soc.* Vol. XXXVI. p. 299. 1910.

24. ERNST, A. The New Flora of the Volcanic Island of Krakatau. Translated by A. C. Seward. Cambridge, 1908.

25. PRAEGER, R. L. A Tourist's Flora of the West of Ireland. Dublin, 1909.

26. { RENDLE, A. B. *Sisyrinchium californicum* Dryand. *Journal of Botany*, Vol. XXXIV. p. 494. 1896.
    MARSHALL, E. S. *Sisyrinchium californicum* in Ireland. *Ibid.* p. 366.

27. { REID, CLEMENT. The Origin of the British Flora. London, 1899.
    REID, CLEMENT and ELEANOR M. On the Pre-glacial Flora of Britain *Journ. Linn. Soc.* Vol. XXVIII. p. 206. 1908.

28. LANKESTER, SIR EDWIN RAY. Nature and Man. The Romanes Lecture. Oxford, 1905.

29. HUXLEY, T. H. On a Piece of Chalk. *Collected Essays*, Vol. VIII. London, 1896.

30. JUKES-BROWNE, A. J. The Building of the British Isles. London, 1911.

31. RAY, J. Three Physico-Theological Discourses, etc. (2nd Edition.) London, 1693.

32. WOODWARD, J. An Essay toward a Natural History of the Earth. London, 1695.

33. SOLLAS, W. J. The Age of the Earth, and other geological studies. London, 1905.

34. WARD, L. F.   Status of the Mesozoic Floras of the United States. *Monograph* 48, *U. S. Geol. Surv.* 1905.

35. ARBER, E. A. NEWELL.   *Cupressinoxylon Hookeri* sp. nov., a large silicified tree from Tasmania. *Geological Magazine*, Vol. I. [v.], p. 7.   1904.

36. { 1. HOLMES, W. H.   Fossil Forests of the Volcanic Tertiary Formations of the Yellowstone National Park. *Ann. Rep. Geol. and Geogr. Surv.* (U.S.A.), 1878, Pt. II. p. 47.
2. KNOWLTON, F. H.   Fossil Flora of the Yellowstone National Park. *Monograph* 32, *U. S. Geol. Survey*, Pt. II.   1899.

37. DARWIN, C.   Journal of Researches into the Natural History and Geology of the countries visited during the voyage round the world of H.M.S. 'Beagle.'   London, 1902.

38. DUFFERIN, LORD.   Letters from High Latitudes.   London. N.D.

39. AMUNDSEN, R.   The North-West Passage.   2 Vols.   London, 1908.

40. MACDOUGAL, D. T.   Botanical Explorations in the Southwest. *Journ. New York Botanical Garden*, Vol. v. p. 89.   1904.

41. { GOEPPERT, H. R. and A. MENGE.   Die Flora des Bernsteins. Danzig, 1883.
CONWENTZ, H.   Monographie der baltischen Bernsteinbäume.   Danzig, 1890.

42. HOVELACQUE, M.   Sur la Nature végétale de l'*Aachenosaurus multidens*. *Bull. Soc. Belge de Géol.* etc., Tome IV. p. 59.   1890.

43. OLIVER, F. W. and D. H. SCOTT.   On the structure of the Palaeozoic seed *Lagenostoma Lomaxi*. *Phil. Trans. R. Soc.* Vol. CXCVII. p. 193.   1904.

44. KIDSTON, R.   On the fructification of *Neuropteris heterophylla*. *Phil. Trans. Royal Soc. London*, Vol. CXCVII. p. 1.

138   BIBLIOGRAPHY

45. SMITH, GEOFFREY.   A Naturalist in Tasmania. Oxford,
        1909.
46. CAMPBELL, D. H.   On the Distribution of the Hepaticae,
        and its significance. *New Phytologist*, Vol. VI. p. 203.
        1907.
47. KIDSTON, R. and D. T. GWYNNE-VAUGHAN.   On the Fossil
        Osmundaceae. *Phil. Trans. R. Soc. Edinburgh*, Vols.
        XLV., XLVI. 1907–09.
48. SEWARD, A. C.   Fossil Plants. Vol. I. Cambridge, 1898.
49. WALLACE, A. R.   The Malay Archipelago. London, 1886.
50. REID, C. and ELEANOR M.   The Lignite of Bovey Tracey.
        Phil. Trans. Royal Soc. London, Vol. 201, p. 161. 1910.
51. HOLLICK, A. and E. C. JEFFREY.   Studies of Cretaceous
        Coniferous remains from Kreischerville, New York.
        Mem. New York Bot. Garden, Vol. III. 1909.
52. HUTTON, F. W.   The Lessons of Evolution. London, 1902.

INDEX

For EU product safety concerns, contact us at Calle de José Abascal, 56–1°,
28003 Madrid, Spain or eugpsr@cambridge.org.

www.ingramcontent.com/pod-product-compliance
Ingram Content Group UK Ltd.
Pitfield, Milton Keynes, MK11 3LW, UK
UKHW010851090126
466816UK00011B/153